Consumer and Sensory Evaluation Techniques

Consumer and Sensory Evaluation Techniques

How to Sense Successful Products

Cecilia Y. Saint-Denis
Westfield, New Jersey

Registered Office(s)
John Wiley & Sons, Inc., 111 River Street, Hoboken, NJ 07030, USA
John Wiley & Sons Ltd, The Atrium, Southern Gate, Chichester, West Sussex, PO19 8SQ, UK

Editorial Office
9600 Garsington Road, Oxford, OX4 2DQ, UK

For details of our global editorial offices, customer services, and more information about Wiley products visit us at www.wiley.com.

Wiley also publishes its books in a variety of electronic formats and by print-on-demand. Some content that appears in standard print versions of this book may not be available in other formats.

Library of Congress Cataloging-in-Publication Data

Names: Saint-Denis, Cecilia Y., 1972– author.
Title: Consumer and sensory evaluation techniques : how to sense successful products / by Cecilia Y. Saint-Denis.
Description: Hoboken, NJ : John Wiley & Sons, 2018. |
    Includes bibliographical references and index. |
Identifiers: LCCN 2017041389 (print) | LCCN 2017052695 (ebook) |
    ISBN 9781119405603 (pdf) | ISBN 9781119405573 (epub) | ISBN 9781119405542 (pbk.)
Subjects: LCSH: Sensory evaluation. | New products–Evaluation. | Marketing research.
Classification: LCC TA418.5 (ebook) | LCC TA418.5 .S35 2018 (print) | DDC 658.5/752–dc23
LC record available at https://lccn.loc.gov/2017041389

Cover Design: Wiley
Cover Image: © Jacqueline N. Denham

Set in 10/12pt Warnock by SPi Global, Pondicherry, India
Printed and bound in Malaysia by Vivar Printing Sdn Bhd

10 9 8 7 6 5 4 3 2 1

"Even the biggest most competent companies fail. The trick is to create an organizational culture that accepts failure so that you can fail small ... rather than failing big."

"True innovation requires learning from the complexities of each failure – a skill that most companies fail to hone."

Samuel West, Museum of Failure,
Helsingborg, Sweden

Contents

Preface

On a Sunday morning, the doorbell rings. UPS just threw on my porch a box that contains the precious item I ordered online less than 24h ago. It is the latest technological tool that everybody wants. Its reviews, which I read thoroughly, are outstanding, five stars across the board. I hurry to the door, grab the box and open it frantically. Now I have it! Let's see how it works and what new avenues it will open in my daily life. At the same time, my son, a 4-year old toddler, grabs the cardboard protecting the device and turns it into a spaceship, which will keep him entertained for a couple hours.

At first sight, the new device is appealing, with a tasteful and modern design, luxury colour and lines, smooth and pleasant texture to the touch. When I get to the instructions, they seem less intuitive than expected. As the new generation shopper that I am, I want to be able to operate immediately. I do not conceptualize it but the on/off bottom seems a little cheap. It might be the noise it makes or just the material itself. I persevere for a few hours. Ultimately, this tool doesn't revolutionize my life. Some of its functionalities are redundant with my previous device. It's not worth the effort. I might use it but it will eventually die as quickly as it emerged.

Meanwhile, my son has made the most out of the cardboard, which, after being a spaceship became an artistic and colourful fort and a car track. But its appeal has died as well, leaving space for new outbursts of imagination.

Nowadays, consumers crave for unique, authentic, customized products. Consumers actively search and seek everywhere rather than passively responding to advertising. In that context, big consumer packaged-goods companies struggle to sell their products. The market is inundated with a never-ending variety of offers making it more challenging to be visible and leaving very little room for innovation. For small, as well as for giant companies, the motto seems to be 'innovate or die'. However, launching a real breakthrough innovation has become a hard-to-achieve and hard-to-predict holy grail.

For years, in the pre-social media, pre-Amazon era, demand for innovation was lower. Big companies could get distributed much more easily than smaller ones, and consumers were used to seeing the same products on the market.

Today, consumers are more informed, more aware and have wider access to very small brands and sellers. Through cooperative websites, anybody can create and sell from one side of the world to the other. This causes challenges to brick and mortar traditional stores and major market share points lost by larger brands.

For decades, decisions on innovative products were based primarily on the intuition of a few creators; or on the intimate conviction of a few top managers. Today, the path to innovation has become way riskier. Developing new products, testing them, weighing market response, predicting failure or success has become critical for managers to ensure success and prevent yearly losses. It has become vital for big brands to invest in robust R&D teams and to consolidate their experience to be able to launch great new products and survive.

The challenge is of course to foster creators and creative teams, but first and foremost to be very solid in supporting the creative process to ensure its success. Until today, teams in charge of evaluation methods as to how new products fit into consumers' life have been very pragmatic and worked mostly on an empirical basis. There are no real common manuals on how to systematically approach consumers all along the creative development. Within each company and among the scientific community with the expertise in this domain, knowledge is passed along via word of mouth through a network of connoisseurs. Everybody moves along following their instinct on how best to test and predict. Given the stakes, it is time to issue systematic approaches. This is precisely what this manual is about. Of course, any method and approach will never be carved into stone, as for the following of the creative process, one needs to remain very flexible and open-minded. Each product category, each invention or creation needs to be approached as a unique case. However, a methodological background is necessary to ensure robustness in the process and to circumvent basic pitfalls.

This manual will therefore dive into the global (Chapter 1) and specific (Chapters 2 and 3) aspects of sensory and consumer test designs: how do we test, when, where and with whom. All of it depends on the objectives we want to pursue and the methods we consider. The testing strategies must be developed (Chapter 4) based on where we are in the development process going from a small-scale to a large-scale approach. Very practical elements will be covered such as tools to be incorporated, as well as deliverables and budget. Chapter 5 goes beyond intrinsic product quality with a more holistic picture of real-life market factors. Chapter 6 concludes with considerations to decide whether to outsource studies.

Before diving into the subject matter, I would like to thank the following people for their inspiration, support and for their challenging and curious minds. Everything goes back to them and how generously they taught me at some point in life. I learned to remain open-minded while they instilled in me the desire to always question, grow and learn.

Acknowledgements

Gilles Trystram, General Director at AgroParisTech. During my Ph.D. research and years after, I have kept from him the love for research as a means of always questioning apparent certainties and applying rigour.

Douglas Rutledge, Director of Analytical Chemistry Department at AgroParisTech. Thanks to Douglas, biostatistics have become approachable to me and a true means of rationalizing complex realities such as sensoriality and consumers' minds.

Joseph Hossenlopp defines himself as an independent thinker. His support, advice and guidance have forged in me respect for knowledge, an instinct to always seek for the right answers, as set out by the best specialists, unfolding reasoning, in order to build a new enriched opinion.

Agnès Giboreau, Living Lab at the Institut Paul Bocuse. Agnès rapidly became one of my mentors when we first met in the food industry at the time of the sensory and consumer methods genesis, being deployed in the industry. Agnes is one of those pioneers who extended these methods to all new fields such as the auto industry and now the hotel industry. Her rigor and curiosity of mind stand before me as an immense source of inspiration.

Jacques Barthélémy was the head of the Sensory Evaluation Department at Nestlé until he retired. He left us in the dawn of 2016. Jacques was a pioneer of the implementation of sensory and consumer methods in the food industry when just freshly established in the academia. He fought against all the obstacles as he was convinced of its relevance. Many in our generation grew up and were fashioned in his pugnacity.

Mara Applebaum, AVP Product Performance Evaluation at L'Oreal USA. Mara has been a colleague, a manager and a true mentor all along my journey in the industry. I have learned so much thanks to her immense knowledge, her incredible open-mindedness and desire for permanent innovation and experimentation in our field. Thanks to her very American 'can do' attitude, many of us have learned how to transpose academic guidelines into the pragmatic world.

Annie Hillinger, Partner, Heads Up! Research, Inc. Annie has a very rigorous and pragmatic approach to research in the consumer field. Working with her has been an incredible opportunity to grow and learn from her sense of careful listening, moderation and translation of consumer insight into action and vision for the future.

My former colleagues in the industry. I have had the immense privilege of travelling through food and cosmetic industries. Multiple windows have been opened into infinite fields of application of sensory and consumer methods way beyond the domains where they took off their first steps. All I know, I do owe it to all these people I have met and worked with along my amazing journey.

My family who is my unfailing daily support. My kids, who at times had to endure my professional dedication, have always carried me with their love and recognition. I am grateful to see them grow up with passion, ambitions, aspirations, positive values and critical minds. My husband, who for years encouraged me to share all I have had the privilege of learning and thus gave impetus to this project. My uncle Edgardo Flores-Rivas, former ambassador, who was an unconditional English advisor all along. And finally my friend Jacqueline Denham for designing the beautiful cover.

All these people, for whom I am so grateful, have a common wonderful asset: a unique sensitivity to small sensorial pleasures of life like sharing a sophisticated flavoured home-made meal, while appreciating the subtlest notes enhanced by a rare vintage wine, the sound of a harmonious musical note or the view of a luminous horizon.

1

The Pillars of Good Consumer and Sensory Studies

1.1 Leveraging Existing Consumer Insight Prior to Building a Test Plan: What Do We Already Know?

For a long time, three major departments in companies have taken the lead: Research, Marketing and Sales. A new product was developed and a consequent budget was put in place to push it into consumers' homes. Within the past couple of decades, the media universe that surrounds us has changed in such a way that this simple approach does not suffice anymore. Marketing environment has become way too complex. Companies must deal with hundreds of cable channels, satellite networks and online social media. This makes the interaction between companies, their brands and consumers more complicated and risky (Blackshaw 2008). As explained by Kietzmann et al. (2011) a simple negative post or tweet could turn into a boomerang. At the same time, interaction with consumers has reached a more personal level. This has forced most companies to embrace what is called Consumer Insight in their mind-set and develop *ad hoc* teams within their organization (Stone et al. 2004).

The consumer insight objective is to go way beyond figures and statistics that were traditionally analysed by marketers. Consumer insight research gathers skills from multiple backgrounds: marketers, psychologists and ethnographers. The idea behind it is to get into the consumer's mind understanding what they purchase, why, how this fits into their daily routines, when, as an individual, as a group, as a community. Consumer insight is now the binder that provides understanding on who consumes what and why. Consumer insight digs into geography, seasons, gender, ethnic and cultural background, age differences and the role these factors play in the consumption of each product category. The goal is to find the truth on existing and emerging behaviours, experiences, beliefs, wants and needs. Consumer insight is the tool that allows researchers and marketers to make a new product that coincides with consumer's needs by either finding the appropriate market space for something creators have

Consumer and Sensory Evaluation Techniques: How to Sense Successful Products,
First Edition. Cecilia Y. Saint-Denis.
© 2018 John Wiley & Sons Ltd. Published 2018 by John Wiley & Sons Ltd.

envisioned or by finding the need gap to be fulfilled that steers and inspires creators with new ideas. Consumer insight serves for both push and pull processes (Walsh 1984). Some distinguish the terms insight and foresight. Insight being the ability to interpret present trends to then predict and prepare the future as the foresight approach.

With that in mind, it is easy to understand why, before engaging in the evaluation of any new product, it is crucial to conduct a full consumer insight research on that category. Oddly enough, in many cases, consumer insight gathers information that everyone already knows. It is just a question of putting it together in a meaningful way that will speak by itself and make sense.

After the Canadian writer Coupland (1991) popularized the expression Generation X to designate the individuals that succeeded baby boomers, all other new sociological designations just flowed out naturally for sociologists to segment the different age groups in our societies. Consumer insight often observes behaviours based on this breakdown (e.g. Kumar and Lim 2008). Whether we want to address, for example, Baby Boomer[1] women skin beauty needs, Gen X[2] men soda drinking drivers, Gen Y[3] (Howe and Strauss 2000) reading habits or Gen Z[4] social media activity, the approach will always be the same. What geographical region are we considering? What is the existing market offer? What are the key benchmarks and the more 'niche' players? Is there a seasonal aspect to be taken into account, and what are the trends, the drivers, the needs and the gaps?

In many cases, companies have a lot of information internally into which they can dig before doing any further research (data mining of existing 'primary research' sources is often called 'secondary research'): intrinsic background knowledge, previous small- or large-scale studies done in that category, marketing and sales data are the first sources to consider. Usage & Attitude (U&A) studies are often conducted every 2–5 years for large categories. Very popular in the 1970–1980s these long and costly studies had been misused in the 1990s. However, they quickly came back as a necessity with the drastic evolution of behaviours in multiple categories and with the expansion of many industries towards new emerging markets. Also, running

---

1 Commonly refers to a generation of people born post World War II roughly between 1946 and 1964, now settling into retirement with some level of comforts.

2 Typically covers people born between the mid 1960s and the early 1980s. Relative to the previous generation, Gen X is usually more open to diversity and embrace differences and new technologies.

3 Also referred to as 'Millennials', covers individuals born between the 1980s and the year 2000. Gen Y has been shaped by the technological revolution. They are usually connected and tech-savvy.

4 Generation of children born after the year 2000, also now called 'Centennials'. Today yet to be better analyzed. They are predicted to be constantly connected with high tech communication tools.

those types of studies became easier thanks to online tools. Ultimately, it is always important to confront knowledge and beliefs within the company itself with up-to-date data to avoid *a priori* certainties.

Social media are an immense source of information: blogs, forums, reviews on electronic commerce sites (broad ones like Amazon or Alibaba, or ones more specialized in a certain category), posts on Twitter, YouTube, Pinterest, all the way to public conversations on Facebook or Instagram. Safko (2010) gives a very comprehensive anatomy of modern social media and how they have become an unprecedented and unavoidable window into our society. Depending on the resources the company has, this research, often called 'social listening' can be done internally or externalized to numerous market-research companies who offer the service. Over the past years, several powerful analytic tools and platforms to systematically process the information have been launched on the market, some being free (broadly general such as www. socialmention.com or targeting one single media like Twitter or Google) and some requiring a monthly fee.

The efficacy of them can be assessed in what they measure and how they represent it versus the needs a company has. Many platforms offer online active dashboards and alerts on pre-set keywords (Table 1.1).

Table 1.1 Typical monitored parameters on social media.

Measure	Description
Audience volume	Number of posts, comments, tweets, reviews per unit of time for defined key words on designated media
Audience categories	Definition of who is speaking: gender, age, professional or not, and so on. This is usually assessed through clear identification or languages hints
Audience influence	Passive observers or stronger influencers[a]. Level of influence is now often measured not only by popularity and number of connections of individuals, but also by their forwarding activity with specific algorithms as shown in Romero et al. (2011)
Competitors	Usually assessment of a number of brands mentioned per unit of time in designated media
Sentiments	Positive, negative or neutral connotation of the conversations. This is usually assessed through language systematic analysis by appropriate software or by linguists (Chamlertwat et al. 2012)

[a] Understanding the level of influence certain individuals, groups or formal bloggers (professionals or not) may have becomes a tool that goes beyond consumer insight objectives (Agarwal et al. 2008). Online word of mouth has become extremely powerful. Therefore, it is vital for companies to track it down to head off anything that could be negative or damaging and to empower happy consumers to share to an infinite audience (Blackshaw 2008; Berger 2016), very often now through the influence of an expert authority or a celebrity endorsement.

Online social media are a tremendous resource to understand consumers. However, depending on the subject or target audience, in some cases, information found through them may not be representative enough. Millennials and centennials are undeniably present for most categories. However, if the target audience is Gen X or Boomers, information found may be more partial. Furthermore, depending on socio-economic categories being considered, regions or countries, prevalence of internet and phones may not always ensure total representativeness if research is only done via online social media. Lastly, one must keep in mind that even though people tend to be more and more vocal online, whether they are happy or unhappy with a product or service, human nature does not change much and comments found may more often be on negative experiences (Blackshaw 2008). In such cases, more traditional 'offline' consumer insight research may be considered such as focus groups or ethnographies (Gunter et al. 2002). An extensive methodological description of these is provided in Sections 3.1 and 3.3 of this manual.

There are also many online free resources that allow to investigate market facts and trends such as:

- Google's Marketer's Almanac
- US Census: American FactFinder, County Business Patterns (CBP), Business Dynamics Statistics
- Claritas MyBestSegments by Nielsen

Lastly, another type of secondary research can be done on additional external sources of information such as Pew Research Center (Pewresearch. org) or Mintel (Mintel.com) with their Global New Product Database (GNPD), as well as their Household Market-Research, which has become one of the industry gold standards to access information on new product trends. Although Mintel is very well known, there is an infinity of analytical online panels, tools and programs that offer information, some being broader, some being more specialized: IRI, Symphony Marketing, Ipsos, Dunnhumby. A few platforms that provide consumer insight by tracking new trends and products can be mentioned (non-exhaustive list): TrendWatching (trend watching.com), Euromonitor International (euromonitor.com) or Trend Hunter (trendhunter.com). They usually require monthly fees to be accessible, or offer reports that can be purchased. Of course, there is also a plethora of market-research companies[5] that either have information or can develop

---

5 Several directories can be accessed online:
  directory.esomar.org
  greenbook.org
  marketresearchdirectory.org
  quirks.com
  ama.org

ad hoc studies. It is interesting to consider organizations such as Esomar (esomar.com) which is a global community of researchers and industries on the market-research field. Their publications and seminars are an invaluable up-to-date source on market data and methodology. Organizations such as Esomar usually require an annual membership.

1.2   Pillars of a Test Design

This sub-chapter details the four crucial puzzle pieces (Figure 1.1) that need to be defined to build a test design.

1.2.1   What Are We Testing?

Figure 1.2 shows the four pillars of a test design: product, target, location and timing that will be further developed in details.

1.2.1.1   Circumscribe the Test Product

Before moving forward in defining a test design, it is very important to restrict what is the to-be-tested product. Do we want to test the container, the content (sometimes called the 'juice' in the food or cosmetic industry) or both? One must keep in mind that every single component of a packaged good is going to impact the way the product is perceived overall. Sometimes, a tiny detail can overshadow everything else and determine the overall acceptance, rejection or banality impression. For example, if we are considering testing a yoghurt, when the objective is to test a new recipe, one may consider using bulk and serving it in neutral white or black bowls. It is important to never underestimate how certain components can sway the consumer's perception. Let us imagine that the new yoghurt recipe has a slight beige tone and is presented in a snow-white bowl under regular day light. The beige colour may infer in the consumer's mind a creamier, heavier recipe (see research from Harrar et al. (2011), which

Figure 1.1 The four puzzle pieces of a test design.

Figure 1.2 Product definition.

shows how implicit knowledge on fat content can be based on appearance). In our example of yoghurt in white bowls, it remains an assumption; but the most important fact is that, whatever it infers in consumers' minds, it is often uncontrolled. Colour of the contents is always important to assess prior to launching a test. A very insightful experiment that can be easily done is to have a panel of consumers test a grenadine syrup coloured in green and a mint syrup coloured in red. Under regular day light, there will always be a larger proportion of the panel that will assess the wrong flavour influenced by the colour, compared to a blind test (for instance, under a light that hides the colour). In the same way, one can ask a panel of consumers to rank cocoa flavour intensity of chocolate milks that have been artificially coloured in different intensities of brown not related with the actual flavour intensity. Most consumers will rank on the base of colour intensity or at least be very puzzled by it. These effects have been known and studied extensively over the past decades (e.g. DuBose et al. 1980 or Zampini et al. 2007). Spence (2015) calls this phenomenon 'disconfirmation of expectation' which has been known and studied for a long time as well (Cardello and Sawyer 1992). Disconfirmed expectations tend to have a negative impact. This varies depending on context, culture and age. Indeed, in the same article, Spence shows that people tend to be more open to their expectations being disconfirmed in a fancy restaurant than in a testing laboratory. Some cultures are less open-minded than others with regard to food. Also, several researches, detailed in Spence review, support that children tend to be more open to artificial miscoloured food.

In the cosmetic field, colour of the 'juice' is also very impactful. Some colours intuitively do not naturally match certain categories in the consumer's mind and may cause repulsion. Yet in that aspect, cultural differences can be significant (Madden et al. 2000). A skin care cream will rarely be accepted in anything else than white; however, the variety of colour tones found in the market for that category is wider in Asian countries. Some visual effects like a pearly aspect may infer either luxury, non-natural or any kind of uncontrolled interpretation that can bias the test. Yet not much is known today in that field (Elliot and Maier 2014).

Colour, shape and material of a packaging or container are also aspects that can bias the perception and therefore the results. Here again a multitude of articles have been published supporting this phenomenon, especially in the food domain. Among the recent researches, we can mention Piqueras-Fiszman and Spence (2012), Spence and Wan (2015), Wan et al. (2015) and the state-of-the-art review by Piqueras-Fiszman and Spence (2015). Typically, the industry tends to test in white neutral containers, labelled with minimal information and instructions to limit the impact. Black packaging is also used often. However, Piqueras-Fiszman et al. (2012) have shown that a black plate can tone down flavour intensity in some cases. Also, depending on the material, black can convey a luxury connotation for certain product categories. One must be careful that the colour of the blind packaging never evokes any brand or conceptual universe. In the case of food, a research from Michel et al. (2015) that relates to eatery industry is particularly interesting as it demonstrates that even the type of cutlery used exerted significant impact on food liking. Of course, when consumers know that they are in a testing environment they might naturally relativize. However, in many experiences they do not. Often, tests can be overshadowed negatively by a cheap accessory on which the consumer will focus all his/her attention on (a classic example is the gloves that are put in hair colour kits: if cheap and uncomfortable, they can completely skew the test).

Obviously, the quality of the material that is used (even if just for the test) needs to be inspected conscientiously for its neutrality and for its integrity. As trivial as it may seem, everybody in the industry has one day experienced tremendous loss due to a cheap material that did not function, a nozzle that did not dispense or broke along the testing period. When a product is to be used by consumers for a certain period at home, it is also very important to provide enough product, having previously assessed how much may be needed on a fairly wide range (Section 1.1). For a shampoo or a daily moisturizer, quantities used by women from a same panel may be quite variable. Too often tests are skewed because consumers were upset not to have enough product.

Lastly, although obvious as well, it should be reiterated here that when comparing several formulas or recipes (contents), all should be packaged identically. When comparing a new recipe with an existing market product, the latest should always be repackaged identically to the test product. Too often, competitors are purchased and just covered with tape. Consumers are always intrigued in finding out what they are given and any hint will lead them to assume or find out. Products that are hard to be repackaged could be spray painted. This obviously implies an additional cost but it is worth it. There is a case in which the product should not be repackaged, which is if the package or dispensing device (nozzle for example for a hair spray) is intrinsically linked to the content and therefore overall experience for the consumer should encompass the comparison of the specific packages. In case it is impossible to repackage a shelf product, one should always keep in mind that there can be a bias. Consumers can then be told upfront to disregard the package differences.

1.2.1.2 Do We Test Blind or Identified Products?

Once the product to test is defined, the next question is, how is the product going to be presented to the consumer? Since consumer society has arisen, it has been widely assumed and then thoroughly studied in several product categories, that when a product is presented blind or 'nude', impacting factors are almost exclusively intrinsic (except for possible presentation bias that the previous paragraph shows how to minimize). This means that what is assessed are product physical attributes and qualities; whereas, if the product is presented under any kind of concept or brand, marketing influences become prominent. For years, most companies have gathered knowledge on label, concept, brand impact versus product intrinsic properties. Many are publicly available especially in the food industry. A few examples that can be mentioned are: Allison and Uhl (1964) for the beer category all the way to Shankar et al. (2009) for the confectionery category or Lowengart (2012) for wine. Deliza and MacFie published an interesting review on the subject in (1996). All studies prove that as soon as a non-generic label, a brand, a concept or any kind of context is disclosed, the expectation will be impacted and the qualities of the product will be assessed in relation with that information.

Thus, the main question to be answered is: is the objective of the test to assess how the intrinsic properties of the product are perceived and liked independently of any kind of market context? This is more often the case in early stages of product development or when a product has not yet been positioned on a given market. When this optic is selected, the product should be tested blind with minimal information. This approach will determine more accurately the consumer's true preferences and product true strengths and weaknesses. Those can then be rolled out to the market place within a brand through a specific concept in the appropriate context. Strengths that have been highlighted through a blind test can be emphasized in the marketing communication. When product strengths match the marketing communication programs, it assuredly improves chances of consumers choosing the right product and better guarantees an ultimately increased market share. A fully blind test approach is also necessary when the purpose is to rank a product with respect to other competitive products or prove its superiority in performance or when an existing product is being reformulated, especially if the current product is highly anchored in the market or has a very loyal group of consumers.

Conversely, when more advanced in the development process and more set on the market context in which the product will fall, one might want to present the product in a more realistic way to the consumer. The product holistically presented encompasses how it would be available in real-life with its brand, package, label, and sometimes even price. In some cases, one may want to understand how a formula, prototype, recipe characteristics match a brand or a concept or how it fits in a specific context. In those cases, the test should be designed in a way that answers those questions. Many researchers in the field tend to state that more often than not, tests should not be blind as blind test results are far from the reality of the market and lead to wrong strategic decisions (Raghubir et al. 2008). This should be relativized. It is true that consumers have a complex process in choosing products. Multiple factors will play a role such as physical or chemical properties and ingredients, that determine quality and durability, as well as inter-individual differences (gender, socio-demographic, cultural differences), context, brand and price. Some product categories, such as food or cosmetics may be more impacted than others by consumer senses and how those make them perceive the product. Thus, in categories where senses are predominant in the selection process, blind tests are necessary at one point of the testing strategy to understand the impact of intrinsic properties. For some product categories, brand will be of such impact in defining the product and conveying the quality image overall that it might be crucial to move into non-blind tests earlier in the testing strategy process. Hence, as it will be detailed in Chapter 4, it is very important to define each individual test methodology within a global testing strategy along with the development process timeline.

An alternative that is used quite often is to test the product blind as a first step. And, once the product's intrinsic characteristics have been assessed, present the consumer with a concept and ask questions on concept appeal and product match with the concept (see extract from a blind product assessment questionnaire followed by concept evaluation in Table 1.2). This approach has the advantage of saving time and money to fulfil both objectives. However, one needs to keep in mind that it can give results that are slightly disconnected from reality as the way concept will be judged will necessarily be impacted by the deep focus the consumer had on the product and the product fit assessment will be less spontaneous than in a real situation as the consumer will have internalized all aspects of the product in a much more analytical way than in real-life. Typically, consumer decision-making is non-conscious (Freeman 2000) and most of the time they cannot necessarily explicate the reasons for their experience (Raghubir et al. 2008).

The Appendix of this chapter gives the example of the full questionnaire.

Table 1.2 Extract from blind product questionnaire, followed by concept assessment.

Q1 Overall how satisfied are you with this moisturizing cream?
- ✓ Very satisfied
- ✓ Satisfied
- ✓ Neither satisfied nor dissatisfied
- ✓ Dissatisfied
- ✓ Very dissatisfied

Q2 Overall, how is this moisturizing cream?
- ✓ Better than your usual
- ✓ About the same
- ✓ Not as good as your usual

Q3 What, if anything, did you like about this moisturizing cream?

Q4 What, if anything, did you dislike about this moisturizing cream?

Q5 How likely would you be to re-use this moisturizing cream?
- ✓ I would definitely re-use it (skip Q6)
- ✓ I would probably re-use it (skip Q6)
- ✓ I would probably not re-use it (go to Q6)
- ✓ I would definitely not re-use it (go to Q6)

Q6 Please explain why you would not be likely to re-use this moisturizing cream?

Usage

Q7 Dispensing this moisturizing cream is
- ✓ Very easy
- ✓ Easy
- ✓ Neither easy nor difficult
- ✓ Difficult
- ✓ Very Difficult

(...)

Concept

Presentation of the concept to the consumer: could be via graphics, videos, audio, samples...

QX This concept is appealing
- ✓ Completely agree
- ✓ Agree
- ✓ Neither agree nor disagree
- ✓ Disagree
- ✓ Completely disagree

QY This concept matches the product you just tested
- ✓ Completely agree
- ✓ Agree
- ✓ Neither agree nor disagree
- ✓ Disagree
- ✓ Completely disagree

Also, when the concept is assessed at the end of a blind product test, the questions that can be looked at will be relatively limited as, due to the consumer's fatigue, at the end of a full product evaluation, it would be unrealistic to expect an accurate response to a lengthy concept assessment. Chapter 5 gives more details on how to conduct a detailed concept test.

1.2.1.3   How Is the Product 'Dressed Up': Packaging, Fragrance?

Impact of packaging components (colour, shape, type of dispensing, weight, material) and fragrance characteristics on perception and ultimately liking and preference are commonly recognized today. They have been thoroughly studied in several product categories; many articles can be found in the literature supporting how impactful these aspects are (Gámbaro et al. 2017). Their influence varies depending on product categories. Gatti et al. (2014) shows how packaging (colour and weight in the case of this study), fragrance intensity and expected efficacy can be intertwined in the case of bath soaps.

As stated by Johnson (1997, p.217) or Dixit (2001), for basically any toiletries or household products, fragrance is critical from the moment of purchase (where most buyers will, if they can, open the cap smell and make a purchase decision), all the way through product experience. It can or not inspire first purchase, communicate the product is doing its job, convey positive feelings and ultimately lead to re-purchase and brand loyalty. Many research teams in this domain, especially those working for the fragrance industry, have dived deeper into different components of this phenomenon: how different odours impact feelings (Porcherot et al. 2010) or texture perception (Churchill et al. 2009) and consumer choice overall (Milotic 2003). For some other product categories, it may be impactful but not as much as for the previous one. For make-up and skin care products, consumers may be slightly driven by odour (especially if it is disliked) but are in general much more influenced by efficacy and results. In the food industry, odour is fundamental and varies with many factors, particularly with age (Griep et al. 1997). As opposed to toiletries, it will not always be possible for consumers to assess it during the act of purchase but it will dictate re-purchase. Some call these successive encounters between the product and the consumer the first and second moments of truth. As stated by Löfgren (2005):

> *The first moment of truth is about obtaining customers' attention and communicating the benefits of an offer. The second moment of truth is about providing the tools the customer needs to experience these benefits when using the product. The combination of these two moments of truth makes up the total customer experience.*

In marketing environment, there are also mentions of a 'third moment of truth' when a consumer becomes a fan (or not) of a product and may start spreading worth of mouth comments or engages in social media.

The olfactory universes that are desirable are linked to the product category that is being considered, but they are also deeply impacted by country, cultural differences, age, gender and socio-economic aspects. It is common sense that testing a shampoo in Europe or Asia will require different fragrances to ensure success and many research teams have analyzed and qualified these differences (Ayabe-Kanamura et al. 1998; Chrea et al. 2004; Ferdenzi et al. 2011). Hence, before launching a test in a specific country, it is very important to get closer to local teams and understand what the market is. If there is no knowledge prior to launching a product test or too many uncertainties on how a particular fragrance is going to be perceived, it is highly recommended to proceed with a 'sniff test' to screen fragrance candidates. Then, it is ideally recommended to follow that screening by an in-use fragrance test to confirm best candidate(s) in a real-usage situation. Section 3.2.2 provides a detailed methodology for these two types of preliminary fragrance tests.

Choosing a packaging for a test also requires knowledge on local usage and cultural habits that may vary depending on regions and countries. Variations could be in quantities that are offered, as well as in the type of container (colour/opacity, shape, type of dispensing, weight/size, material). In many European countries, such as France, the typical content of an individual yoghurt is 125 g, Whereas in the United States, an individual yoghurt is more often 150 g. Shampoo or body wash bottles are usually bigger in countries where people wash more frequently. But surprisingly as well, countries where people travel a lot are now very used to small one-time dose samples. Hair gels or skin moisturizers are most often presented in jars in Europe, whereas is countries like Canada or the United States, gels usually come in tubes and moisturizers will be found in pump bottles. Indeed, consumers in these countries do not like the idea of 'scooping' with their fingers for hygiene reasons. Conversely, countries in Latin America are used to big family size transparent gel jars, as the habit is for the whole household to share the same product (boys will hold their short hair style and girls will hold fly-aways on their ponytail with the same gel). Laundry detergent in small powder packets is still very common in countries where prevalence of hand wash is high compared to big liquid laundry detergent bottles or packs of liquid laundry pods that have become popular in some other countries. As obvious as it sounds, prior to producing samples for a test, it is necessary to gather that very specific consumer insight on the most appropriate way to present them, unless the objective is of course to introduce something totally new and assess acceptability. In the latter, it is still necessary to be fully aware of the established market.

Lastly, depending on climate and season, it is always important to check if the packaging is appropriate in terms of stability of the product overall and of fragrance, in particular.

1.2.1.4   Experimental Design: Order of Product Presentation

When multiple products are to be tested, it is commonly established that the order of product presentation needs to be randomized among all subjects to limit the presentation order or carry-over effects. Thus, if a test is done on three products: A, B and C, the six combinations ABC, ACB, BAC, BCA, CAB and CBA need to be seen the same number of times (whether products are given simultaneously, in which case they are to be displayed in that order, or sequentially). Therefore, in this example, the number of subjects needs to be a multiple of 6 to have a perfect balance in the design. This is true across all types of tests, sensory experts' tests, as well as consumer qualitative and quantitative tests.

It is always better to have all subjects test all products (**Complete Block Designs CBD**). However, that is not always feasible in practice. The maximum number of products that one subject can assess per session, or over a given time frame, needs to be cautiously determined based on potential saturation or fatigue, logistics, timing or budget constraints. For some products, like coffee or alcoholic beverages for instance, although protocols may state that products are not swallowed (especially for sensory), one can easily understand how it is impossible to have one person assess more than a few products at a time. In some cases, the limit may be purely due to logistics: for hair products like shampoos, it is impossible to assess more than one at a time when the product is self-applied. In the latter, multiple products can be evaluated by the same subject but more time is required. In the case of fragrance tests or very fragrant products (hairsprays for instance), one must be very cautious when there are multiple products assessed at a time, to give the subjects enough breaks to reset and to have a proper ventilation. Sometimes, the limit is based on the quantity of products available. For some luxury products, cost can be a limit as to how many samples are available for the test. In other cases, sourcing enough products can be challenging when considering competitors. Online ordering capabilities nowadays may ease that aspect, but availability and discretion that one may require makes large-scale purchases not always easy to achieve.

An alternative, when it is impossible to have all subjects assess all products, is to fall back on **Balanced Incomplete Block Designs (BIBD)** (Wakeling & MacFie 1995; Ball 1997; Hinkelmann & Kempthorne 2005).

To ensure that an incomplete block design is balanced, if s is the number of subjects and p the number of products (Figure 1.3), the design needs to be built considering the following rules[6] (Khare & Federer 1981):

- Each subject sees the same number of products x $(x < p)$
- Each product is seen the same number of times y in total $(y < s)$
- All product combinations (pairs) are seen the same number of times λ

---

6 Most statistical packages, such as SPSS, SAS and R offer modules to build balanced incomplete block designs in which all order effects are automatically randomized and balanced.

From the complete design, if for example only three out of four products can be seen per subject (the second chart of Figure 1.3 does not consider additional randomization of presentation order that is needed to fully balance the design), for each row, six combinations are to be added.

In the case of large-scale quantitative consumer tests, when products are to be used at home and not in a central location test (CLT), they can either be given all at the same time with clear instructions on the order in which they need to be used (in that case there is always a risk consumers may not follow it), or consumers can come pick up the following product each time they are done using one and bringing it back. The order in which products are given is randomized as well and overall the experimental design can be a CBD or a BIBC. The term commonly used to name this type of large-scale consumer test is a **sequential monadic test**. Some use the term of **cross-over design** (Figure 1.4). This means that consumers have one product at a time, but overall, they are going to test several products one after the other in a randomized order. The block of products they test and the

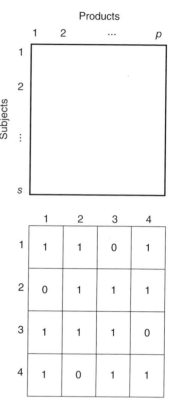

Figure 1.3 Illustration of balanced incomplete block designs.

sequence in which they test them, will necessarily impact their results. But if the design is well balanced, the effects will be smoothed out. A given product will have the biggest impact on the one that follows right after as the consumer will consciously or unconsciously compare certain aspects. The impact on any subsequent product after the one that comes right after will depend on the length of the testing period for each single product. It is important to rule out cases where some very unique or outstanding products may impact the whole sequence.

There are situations where certain constraints impose that each consumer will only be able to assess one product (e.g. we want to compare three shampoos, but each one needs to be used over a period of several weeks and months of field is not an acceptable option; one of the products is so unique it may bias the whole series on a sequential monadic). In such cases, it is common practice to test each of the products on a large group, called a monadic cell, and then to

Subject	Product	Order[a]	Q1	Q2	...	Qn
1	A	1				
1	B	2				
2	A	2				
2	B	1				
⋮						
N	A	1				
N	B	2				

[a]Number of times A and B seen in first and second positions are equal.

Figure 1.4 Sequential monadic with *N* subjects and two products, Q1–Qn questions.

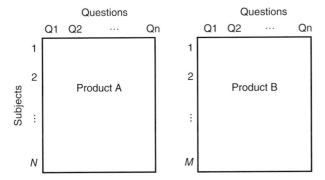

Figure 1.5 Monadic cells with two separate cells for products A and B (N and M subjects, respectively).

compare results of the different **monadic cells** (Figure 1.5) that are done in parallel. To compare large-scale monadic cells rigorously, it is of primary importance to just change one parameter between cells. Volunteers screening characteristics need to be strictly the same (see Section 1.2.2.2 on recruitment criteria) as well as regions, climate and timing. It is just the product that will differ. It is strongly discouraged to compare results from a cell run in winter with results from a cell run in summer for example. Hence, if the design is rigorous, it will be possible to designate which product performs the best by comparing the cell results. Size of the cells typically needs to be of 200+ consumers, and in order to compare two cells, their sizes need to be comparable.

Monadic cell approach versus sequential monadic tests to compare products is the object of many debates in the industry.

- **Monadic cell** simulates more of a real-life situation and results have the advantage of providing absolute values not biased by a direct comparison on all measures (hedonic, intensity, efficacy, just-about-right questions). It also has the advantage that if a test is conducted and a few months later a new product needs to be compared with the ones initially tested, technically a new cell can be run if climate, timing and recruitment criteria are the same. However, there can always be a bias linked to the variability of the respondents between cells. It requires a rigorous and identical recruitment for the multiple cells.
- **Sequential monadic** will be preferred when a direct comparison is required and an answer to an overall preference question is expected. The comparison is more precise as the variability of the respondents is controlled. In some cases, first position monadics, in a sequential monadic, are compared to obtain more absolute values. It is however risky if the sample size is not large enough (see Section 3.2.4).

In practice, the choice often depends on logistic constraints. Furthermore, it is common practice in the industry to use sequential monadic approaches when the project is sufficiently advanced so that a quantitative test is being considered, but still early in the innovation stage. In those cases, a sequential monadic test will allow to explore how several products compare (internal options vs external competitors) and give guidance to the development teams. Typically, when projects are more advanced, large monadic cells are used for validation. They are also commonly used for benchmarking. Monadic cells are the norm to compare results on a same product between different populations for example. Here again, one needs to be very cautious in what parameters differ between the different cells that are being compared.

1.2.2   With Whom Are We Testing?

The target definition is one of the key aspects of a test design. Figure 1.6 highlights the different elements that are described in this section for that purpose.

1.2.2.1   Who Are the Competitors and Benchmarks?

As described in Section 1.1, consumer insight will provide information on main competitors, quantify their market share, and the eventual threat they represent. Based on that, unless there is no equivalent on the market, one must define who the benchmark(s) are for the product that is being developed (they can be external or internal sometimes). Often benchmarks are defined based on their market share in value or units sold. However, a benchmark is not necessarily the number one on the market for a given category but a product

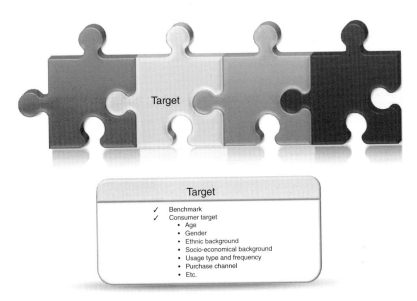

Figure 1.6 Target definition.

that has similar properties or benefits or a similar market target. For example, if a product with natural ingredients is being developed, the benchmark for the test is not necessarily going to be the market leader, that is not positioned in that 'natural ingredient' segment, but emerging products in the category that claim similar merits. A benchmark in some cases may be a very 'niche' brand, if the objective is to hit that very small segment to make it emerge and grow.

Defining the benchmark(s) is a key step as it is the basis of many consecutive decisions. One of them is the consumer target. In a study where there is a clear benchmark, it is strongly recommended to target consumers that eat, drink or use that reference product to assess how the new proposal is positioned versus the state of the market.

In some product categories, it will be very easy to recruit regular or even exclusive users of a well-defined brand. In some others, it might be more challenging, either because the market is small and both time and budget to find those consumers will become exorbitant, or because consumers are not loyal to a specific brand and might switch easily over time or consume several brands simultaneously. When it is anticipated that a recruitment will be challenging, it can be decided to try for a defined period (based on cost and deadlines) to recruit on ideal criteria and then eventually to open to a wider range of brands. In some situations, it might be decided right away to include a larger number of brands in the target definition. However, it cannot be too broad; otherwise it might be very difficult to interpret the results when consumers are comparing to usual. In case it is impossible within time and budget to recruit the ideal users, an alternative can be to include the benchmark in the test (usually the case in blind tests). It is very important in the recruitment screener to be extremely specific on the benchmark reference that is being used; many brands have myriads of variations under the same name. Sometimes the same reference can have different strengths, different fragrances or different colours. It is also crucial to ensure the competitor has not relaunched or reformulated their reference. Therefore, the screener needs to be precise to avoid inclusion errors. If the recruitment is done face to face or online, it is recommended to add unambiguous pictures. If the recruitment is done over the phone, the recruiter can still have the pictures at hand and ensure via straightforward questioning that consumers are referring to the right product.

Lastly, it can be important, depending on the objectives of the study, to add questions regarding the benchmark, such as:

- What is your opinion on your current product (on a scale from like very much to dislike very much), if the objective is to recruit satisfied users?
- How often do you use it? We may want to make sure the usage is not extremely occasional.
- Which product is your 'go to' product (especially if recruitment was opened to several usual products?
- How do you use/consume the product? Some products might be used differently and the study may be targeting specific users. For example, a Greek yoghurt can be consumed during breakfast or as a snack, a face moisturizer may be used in the morning or in the evening, a hair product may be used to achieve different styles, a software or an electronic device may be used only for very specific modules or tasks.

1.2.2.2 Who Is the Target (Age, Gender, Socio-Economic Background, Users of and so Forth)?

Defining the target of people that are going to be included in the study is one of the foundations that ensure exploitability of results. The set of inclusion and exclusion criteria are what is commonly called the screener. It usually consists of a series of multiple-choice closed-ended questions. The accuracy of the recruitment relies ultimately on the fact that people who end up being included, as a result of the screening process, correspond to the defined target. Recruitment can be done in person, over the phone or online. Therefore, inclusion can be based on the recruiter's judgement (if in person or over the phone) or based on self-declared characteristics. To avoid errors, misinterpretations or false declarations, the multiple responses that are available will encompass broad enough choices that the consumer will not guess or assume which responses may include him/her or not (in many cases consumers are eager to be included).

Table 1.3 details attributes that are usually incorporated in a screener to define a target.

Table 1.3 Screener criteria.

Criteria	Examples of breakdowns for questionnaire
Age	Less than 18; 18–24; 25–34; 35–44; 45–54; 55–64; 65 or more
Gender	Male/female
Ethnic background[a]	Caucasian; Hispanic/Latino; Black/African; Asian; Other; Refuse to answer
Socio-economic background	Income brackets (variable per country and currency)
Occupation[b,c]	Advertising or public relations; journalism; marketing/market-research; bank; insurance; medical doctor/pharmacist; food industry, beauty/hygiene/personal care industry, other
Current or recent participation in a consumer panel[d]	Yes/no
Current or recent participation in a test in the same product category	Yes/no
Currently pregnant[e]	Yes/no
Product usage	Depending on the study, this will encompass a list of sodas, snacks, dairy products, hair products, make-up products, laundry detergents, and so on for the respondent to choose (all that applies) and for the recruiter to be able to select people using the appropriate products within the category
Brand	This will encompass a comprehensive list of brands within the category to select people using the target brands (if applicable)
Purchase channel[f]	Discounted department stores (Walmart, Target, etc.); supermarket; beauty supply stores, wholesale clubs (Costco, Sam's, etc.); drugstores; Internet; department stores; other
Frequency, type of usage	As stated in Section 1.2.2.1, we may want to target frequent or 'heavy' users of the category (depending on the product type, frequencies to call someone a frequent user are variables), or occasional users, first-time users, abandoners (this can be interesting to understand people who have experienced the category but left it, the reasons for that and if they may re-enter the category). We may also want to target people who use the product in a specific way

(Continued)

Table 1.3 (Continued)

Criteria	Examples of breakdowns for questionnaire
Additional information depending on product categories	For hair products[g]: selection may be based on hair demographics (length, thickness, abundance, curliness, colour, frequency of usage of certain treatments, etc.)
	For skin products and make-up: selection may be based on skin characteristics (tone, dryness, wrinkles, current shade usage, etc.)
	For general focus groups, selection may be on currently encountered problems, expectations or needs

[a] In some countries, this criterion may not be included or stated as 'country of origin'.

[b] In most studies, it is recommended to exclude people working in advertising, journalism, marketing. One might also want to exclude people working in the same industry as the test product.

[c] Usually the question is broad and encompasses the respondent's family and close friends.

[d] In certain studies, it can be recommended to exclude people who are already participating in a study or if their last study was very recent (typically 3–6 months), especially if the study involves products in a similar category. The main reason is to avoid too 'professionalized' consumers and ensure more spontaneous responses. Also for some products, if the consumer has any type of adverse reaction, it is better not to have any overlap between multiple products being tested to have better chances to identify whether the origin involves the product being tested.

[e] It is always necessary to determine whether we want to exclude pregnant women, especially when there is a product interaction that might be sensitive.

[f] Some product categories may have very specific channels (hair products may be purchased in hair salons per Hairstylist's advice, personal care products may be purchased at the aesthetician's). Also, each country may have their specific channels. For example, many countries have an important door-to-door market for certain categories; in certain states or countries, wine and liquors can only be purchased in liquor stores.

[g] To screen participants on hair or skin characteristics; in some cases shade charts or pictures may be sufficient to support recruiter's work. In some cases, a specialist may be needed, such as a hairdresser, an aesthetician or even a medical doctor depending on the objective of the study.

The appendix of this chapter gives an example of a complete screener.

One very important aspect to consider to ensure that the screening process will be efficient (time and cost optimization) and successful (ultimately the right persons are going to be enrolled in the study) is to fully understand motivations of consumers to participate. Hennig-Thurau et al. (2004) gave an interesting review on what motivates consumers to articulate themselves (this study was specifically looking at situations where surveys are carried online but many aspects are true for in-person studies). Of course, economic incentive is one of the main reasons. However, interestingly, consumers are also eager for social interaction, they want to share their concerns or knowledge with other consumers or peers and they also seek to grow their self-esteem. When studies look for participants based on their expertise on a given subject, valorizing that expertise during the recruitment discussions may trigger willingness to participate, sometimes even allowing to lower the incentive. In some cases, incentives could be coupons, gift cards or products, provided the participants see value in them.

Lastly, it is important to know the countries' rules governing the taxation of incentives, be transparent upfront with the participants and collect all their information for tax purposes during enrolment. It is also necessary to always obtain written consent forms for each data collection activity with participants. A consent form to be signed and dated by the participants needs to clearly explain the project, what is expected from the participant in terms of process, information recorded, pictures, videos and what will be done with those, including confidentiality aspects. Benefits and eventual risks need to be stated as well as length of the study and conditions to withdraw. Through the informed consent form, it is very important to ensure that people understand what it means to participate in a study so they can decide in a conscious and deliberate way to be enrolled or not. The appendix of this chapter gives an example of a consent form.

This section has as its sole objective to describe rules to recruit for consumer studies (small or large scale), for which not much literature can be found, but it does not cover the rules governing clinical trials. If more information is required in that domain, one can refer to one of the many comprehensive manuals such as the one written by Pocock (2010). Section 3.3 describes selection of patients; many elements are of interest for consumer studies as well. More recent manuals from Friedman et al. (2015) or Brody (2016) are also enlightening as they detail up-to-date ethical, legal and regulatory aspects.

1.2.3 Where Are We Testing?

Figure 1.7 highlights the third pillar of a successful test design is a clear definition of the location. This is thoroughly described in the subsections below.

1.2.3.1 Circumscribe the Geographical Region or Country

It is common sense that, most often, results from a study carried in one region cannot be extrapolated to another region, within the same country and *a fortiori* worldwide. Quite often, within the same region, depending on whether the study is conducted in a specific inner city or in its suburbs, results will vary as people do not have the same lifestyles, activities or professions. We can take the example of Florida State in the United States, known for having a large Hispanic population. Therefore, when running a study wherein the target is Hispanic consumers, one may consider Florida. If we look at the United States Census Bureau results (census.gov), Hispanic/Latino Populations are predominantly located in the South where some areas show up to 67% of the population

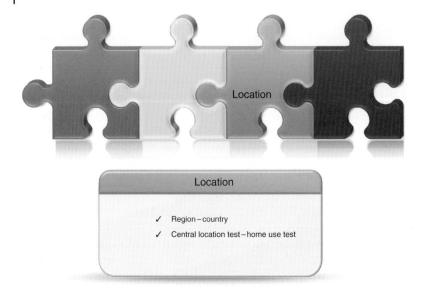

Figure 1.7 Location definition.

as being Hispanic/Latino as opposed to the North where some areas have less than 3% Hispanics. Census information is an extremely important source to make decisions on where to run a study. Figures and maps publicly available on their official website give very precise information on how the population is distributed geographically in terms of age, race (for certain countries only), education, employment, type of housing and size of households or spoken languages. Nowadays, most countries have census information available that it is essential to consult. Most government websites provide very up-to-date information with interactive maps or dashboards that are pretty user-friendly and even pleasant to play with.[7]

Depending on how broad the recruitment screener is, targeting the right region has different implications. For instance, if a product is not *a priori* targeting a group from a specific ethnic background or socio-economic range, understanding the dichotomy of the zone will allow to have a clear photography of the panel that is being portrayed. Then, it will be possible to ask for the appropriate questions to get a relevant profile of the consumers, ideally balance some factors, and cross-check which have any impact on results. Conversely, if

---

7 Few examples:
Insee.fr in France
inegi.org.mx in Mexico
ibge.gov.br in Brazil
censusindia.gov.in in India

a specific product is intended for a more defined consumer group, running the study in the most appropriate zone will make the recruitment process easier and more cost-efficient. Going back to our example, when looking for Hispanic/Latino consumers to test a very specific type of food recipe, it is better to run the study in the southern areas of Florida. Pushing things even further, it is important to know that Hispanic/Latinos living in Florida are predominantly Cuban, whereas those leaving in Los Angeles area are predominantly of Mexican descent, and those living in the North-East coast are more likely to be from Central America countries. Their cultural backgrounds are not interchangeable and may have profound specificities that can be critical or not, depending on the product category that is being considered. Their origin will impact their habits, needs and expectations.

1.2.3.2   What Is the Impact of Local Culture?

As stated by Moran et al. (2014), prior to entering any market, it is vital to 'spend time and money on *due diligence*'. On page 12 of this manual, 10 key dimensions are listed as impacted by culture and needing to be assessed in any due diligence.

1) Sense of self and space
2) Communication and language
3) Dress and appearance
4) Food and feeding habits
5) Time and time consciousness
6) Relationships
7) Values and norms
8) Beliefs and attitudes
9) Mental process and learning
10) Work habits and practices

These dimensions, defined as inter-correlated, seem fundamental to analyze when setting up a consumer study. Having them in mind allows to understand implications of testing the same product in one region versus another region. Let us picture the following scenario: on a warm and humid morning in Rio, Brazil, there starts a testing sessions for personal care products among women in their forties. Anybody who has experienced this knows that in no time these women will become friends, exchange phone numbers, discuss their children, passions, hobbies and beauty tips. The atmosphere will be jovial and rather loud. For somebody running a focus group, difficulty will reside in channelling the discussions rather than making sure people expand. For somebody who is collecting individual opinions, the challenge will be to limit point-of-view exchanges. But in a similar situation in Tokyo, Japan, women will be astonishingly punctual, and the atmosphere will be silent. Panellists will not spontaneously socialize nor exchange opinions. Moderators must be knowledgeable of language, communication codes and sensibilities for a completely different cultural environment. Running tests in countries such as Japan may present the difficulty that volunteers are less spontaneously inclined to express discontent as it may be considered disrespectful. Inversely, some cultures tend to be negative more easily, some others will fall into excessive or artificial enthusiasm in a test environment which is ultimately misleading.

Scales will be discussed in detail in Section 3.2.2. Regarding cultural differences, it is important to mention that offering a scale with middle points (and therefore even numbered response categories) produces different reactions in different cultures (Si and Cullen 1998); this is called the 'central tendency' which is more pronounced for some. Some cultures are also more cautious than others in using extremes in a rating scale and will always be more nuanced (Chun et al. 1974). However, this tendency will also depend on the type and length of the scale that is offered. Hui and Triandis (1989) showed in their specific study how Hispanics have a greater propensity to use extremes. This, however, can be reduced when using 10-point scales as compared to 5-point scales. Most of the literature that can be found on scales and cultural differences relates to surveys done in the work place encompassing employees from different origins or done within a same company in different countries. Indeed, a lot of research on cultural differences in the work place has been deployed over the past decades. Some articles also show how age and education influence scale usage (Stening and Everett 2010). Some interesting literature can also be found that relates to political surveys which shows usage of extremes in scales varies depending on cultural or ethnic origin (Bachman and O'Malley 1984). Yu et al. (2003) give an analysis of central tendency across cultures for semantic differential scales used in Market-Research. In all cases, before launching a study in a specific region, it is very important to get closer to local teams (internally, local recruiters or local market-research companies) to ensure relevance of scales and vocabulary that will be used, especially when translations are involved. Most often, locally based teams have a very valuable field knowledge which must be exploited rather than transferring a global model.

1.2.3.3 Do We Test In-Home or in a Central Location?

Very often, it is preferred to test products within the development laboratory facilities or in a Central Location Testing area (commonly designated by CLT) rather than at the tester's home (home-use test (HUT), or in-home-use test (iHUT)) mainly for cost reasons. Some studies have compared results and shown no significant effects on hedonic ratings for the products being tested (Hersleth et al. 2004). However, others (Boutrolle et al. 2007) have shown that it depends on the products and on the way and context they are usually eaten primarily (within a family meal or a social event or in a lonesome context).

Thus, the quality of the prediction of real-life environment results may depend on the product that is being considered. Boutrolle et al. (2005) showed that in the case of the fermented milk beverages that they tested, liking scores are higher in the HUT. Thus, if a dairy product is consumed as a stand-alone snack or as a dessert, in an authentic spontaneous situation or not, results can be different. For beauty products, if a woman tests a product as a stand-alone in a CLT, results can be quite far from reality compared to the product being part of a complex multi-product routine in her own environment. Often, for a woman to participate in certain tests, she may be asked to comply with rigid constraints (e.g. please come with your face washed with, no cream, no make-up). This woman is thus placed in an artificial setting, away from her own mindset (whether she may comply or not, something that often happens and may not be visible but will still impact her own judgement).

On the other hand, giving the product to be used at home may imply a higher number of random and uncontrollable factors. Indeed, Boutrolle et al. (2005) has also shown that CLT yields more robust results than HUT. The water used to prepare a coffee or to rinse a shampoo will have variable mineral content, the light used to assess a blush or an eye-shadow result will be variable, the bathroom ventilation in which a fragranced product is applied may not be efficient. It is also impossible to control whether the product will be exclusively used by the person who was recruited. Other people in the same household may try it and thus influence the final judgement.

All this being considered, it has been a typical practice to fine tune formulas and guide product development via more controlled CLTs as a first step, and then move onto larger-scale HUTs when the development teams feel the product is more robust and ready (Griffin and Stauffer 1990). Chapter 4 gives more clues on this path and when it is clever to move to the next step.

Another aspect to keep in mind, when deciding on whether to test in a CLT or in an HUT, is if the product requires the consumer to use it multiple times to enable her or him to express an opinion. For some products, especially if they are very innovative, typically the first encounter will be the opportunity to discover and gauge. The second time will allow some adjustments (instructions, if any, become clearer, consumer plays with quantities to reach the optimal ones). The third time, the consumer arrives at a more accurate opinion. For some products, the need for multiple uses may be dictated by the fact that the product can build up or create some fatigue, weariness or disgust, despite an initial enthusiasm. In cases where it is anticipated that the product may require multiple usages, it will be more cost-efficient to conduct a HUT.

Finally, two points need to be kept in mind. Many companies conduct tests in their own facilities. It is important to know that this impacts impartiality, as the participants know where they are and might be more inclined to complaisance

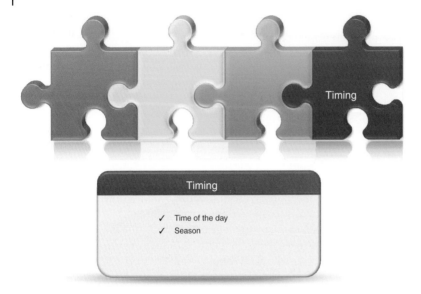

Figure 1.8 Timing definition.

or exacerbated criticism. Also, for some products, there might be a patent pending or some confidentiality considerations that might restrict the possibility to let the product go to people's homes.

1.2.4   When Are We Testing?

Lastly, Timing of a test is the last key pillar of an efficient test design as highlighted in Figure 1.8.

1.2.4.1   How Important Are Consumer Habits?

Common sense is the best indicator as to when is the best moment for a product to be tested and it will most often go back to the consumer insight knowledge that has been gathered on that particular category. Too often, products are tested at times, and in contexts, that have nothing to do with real-life. It is true that it is not always possible to completely mimic what the consumer does in terms of timing due to business-hour constraints. However, testing chicken broth in the morning, a night cream in the middle of the day, a heavy body moisturizer in the summer, a hair spray in the Midwest winter while most people are going to be wearing hats, are scenarios that happen quite often and that are evidently not ideal to obtain exploitable and predictive results.

For any given category, consumer habits must be thoroughly analyzed and considered to schedule a test. As much as it is possible, the consumer needs to be put in a setting that mimics his/her habits. In some situations, it is worth shifting business hours to be able to do so. For instance, if we want to test shaving creams on a man's target, the coordinating team must start early in the morning to facilitate recruitment and to enhance significantly the accuracy of results.

1.2.4.2   Is There Any Seasonal Impact?

There is a plethora of products not being used all year long. Some typical examples are sunscreens or ice creams, which are more common in the summer in temperate regions. In terms of food consumption, consumers commonly shift their habits seasonally in regions that experience more drastic climate changes throughout the year. Even though seasonality of crops is not as impactful as it was back in previous days, people still follow certain patterns such as eating more soup, stir-fry and crock-pot recipes in cold seasons. Whereas in warmer seasons, salads and raw vegetables will be more common, as well as consumption of ice cream, frozen yoghurt and cold smoothies. Conversely, in warmer climate countries, hot soup may be a daily habit throughout the year. Many other product categories vary by seasons, such as skin care routines due to seasonal biological changes. Depending on countries and culture, variations can be drastic. Asian women are the most sensitive to changes in their skin biological parameters and therefore have a greater propensity to change their habits (Abe et al. 1980; Youn et al. 2005).

Hence, it is important to choose the right season for each test. If coordinating from a region that has a rough winter, it may be relevant to travel to a more appropriate climate to test certain products.

1.2.5   Target Segmentation Principles: Do We Need to Define Different Consumer Cells?

Segmentation is a widely applied technique to define sub-groups of consumers based on selected parameters. Segmentation can be done *a posteriori*, based on data that has been collected, to group consumers based on assessments which can be verbal material, hedonic ratings or descriptive measures. This will be detailed in Section 3.2.6.

In some instances, the target that is to be addressed may be too broad. In those cases, it is necessary to define multiple cells or segments in the panel that is being recruited, as well as ensure equivalent numbers in each, to enable us to compare results in the end. Consumer segments or cells can be defined by multiple factors, such as gender, age range, ethnicity, socio-economic parameters, purchase channel, product use, loyalty to a product type or a brand, as well as region or season.

Figure 1.9 summarizes the elements described in the chapter.

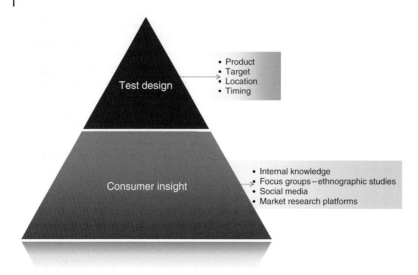

Figure 1.9 Foundation of a successful consumer or sensory study.

References

Abe, T, Mayuzumi, J, Kikuchi, N, Arai, S 1980, Seasonal variations in skin temperature, skin pH, evaporative water loss and skin surface lipid values on human skin, *Chemical and Pharmaceutical Bulletin*, vol. 28, issue 2, pp 387–392.

Agarwal, N, Liu, H, Tang, L, Yu, P 2008, Identifying the influential bloggers in a community, *Proceedings of the 2008 International Conference on Web Search and Data Mining*, Palo Alto, CA, February 11–12, 2008, pp 207–218.

Allison, R, Uhl, K 1964, Influence of beer brand identification on taste perception, *Journal of Marketing Research*, vol. 3, issue 3, pp 36–39.

Ayabe-Kanamura, S, Schicker, I, Laska, M, Hudson, R, Distel, H, Kobayakawa, T, Saito, S 1998, Differences in perception of everyday odors: A Japanese–German cross-cultural study, *Chemical Senses*, vol. 23, issue 1, pp 31–38.

Bachman, J, O'Malley, P 1984, Yea-saying, Nay-saying, and going to extremes: Black-white differences in response styles, *Public Opinion Quarterly*, vol. 48, issue 2, pp 491–509.

Ball, R 1997, Incomplete block designs for the minimisation or order and carry-over effects in sensory analysis, *Food Quality and Preference*, vol. 8, issue 2, pp 111–118.

Berger J, 2016, *Contagious: Why Things Catch on*, Simon & Scuster, New York, NY.

Blackshaw, P 2008, *Satisfied Customers Tell Three Friends, Angry Customers Tell 3,000*, Doubleday Publishing Group a division of Random House Inc., New York, NY.

Boutrolle, I, Arranz, D, Rogeaux, M, Delarue, J 2005, Comparing central location test and home use test results: Application of a new criterion, *Food Quality and Preference*, vol. 16, issue 8, pp 704–713.

Boutrolle, I, Delarue, J, Arranz, D, Rogeaux, M, Köster, E 2007, Central location test vs. home use test: Contrasting results depending on product type, *Food Quality and Preference*, vol. 18, issue 3, pp 490–499.

Brody, T 2016, *Clinical Trials, Study design, Endpoints and Biomarkers, Drug Safety and FDA and ICH Guidelines*, Elsevier, London.

Cardello, A, Sawyer, F, 1992, Effects of disconfirmed consumer expectations on food acceptability, *Journal of Sensory Studies*, vol. 7, issue 4, pp 253–277.

Chamlertwat, W, Bhattarakosol, P, Rungkasiri, T 2012, Discovering consumer insight from Twitter via sentiment analysis, *Journal of Universal Computer Science*, vol. 18, pp 973–992.

Chrea, C, Valentin, D, Sulmont-Rossé, C, Mai, H, Hoang-Nguyen, D, Abdi, H 2004, Culture and odor categorization: Agreement between cultures depends upon the odors, *Food Quality and Preference*, vol. 15, issue 7–8, pp 669–679.

Churchill, A, Meyners, M, Griffiths, L, Bailey, P 2009, The cross-modal effect of fragrance in shampoo: Modifying the perceived feel of both product and hair during and after washing, *Food Quality and Preference*, vol. 20, issue 4, pp 320–328.

Chun, K, Campbell, J, Yoo, J 1974, Extreme response style in cross-cultural research, *Journal of Cross-Cultural Psychology*, vol. 5, issue 4, pp 465–480.

Coupland, D 1991, *Generation X: Tales for an Accelerated Culture*, St. Martin's Press, New York, NY.

Deliza, R, MacFie, H 1996, The generation of sensory expectation by external cues and its effect on sensory perception and hedonic ratings: A review, *Journal of Sensory Studies*, vol. 11, issue 2, pp 103–128.

Dixit, S 2001, Fragrance selection in consumer care products, *Chemical weekly*, November 27, pp 173–175.

DuBose, C, Cardello, A, Maller, O 1980, Effects of colorants and flavorants on identification, perceived flavour intensity, and hedonic quality of fruit-flavoured beverages and cake, *Journal of Food Science*, vol. 45, issue 5, pp 1393–1399.

Elliot, A, Maier, M 2014, Color psychology: Effects of perceiving color on psychological functioning in humans, *Annual Review of Psychology*, vol. 65, pp 1–771.

Ferdenzi, C, Schirmer, A, Roberts, C, Delplanque, S, Porcherot, C, Cayeux, I, Velazco, MI, Sander, D, Scherer, K, Grandjean, D, 2011 Affective dimensions of odor perception: A comparison between Swiss, British and Singaporean populations, *Emotion*, vol. 11(5), pp 1168–1181.

Freeman, W 2000, *How Brains Make Up Their Minds*, Columbia University Press, New York, NY.

Friedman, L, Furberg, C, DeMets, D, Reboussin, D, Granger, C 2015, *Fundamentals of Clinical Trials*, Springer, New York.

Gámbaro, A, Roascio, A, Boinbaser, L, Parente, E 2017, Influence of packaging and product information on consumer perception of cosmetic creams: A case study, *Journal of Sensory Studies*, vol. 32, issue 3, pp e12260.

Gatti, E, Bordegoni, M, Spence, C 2014, Investigating the influence of colour, weight, and fragrance intensity on perception of liquid base soap: An experimental study, *Food Quality and Preference*, vol. 31, pp 56–64.

Griep, M, Mets, T, Massart, D 1997, Different effects of flavour amplification of nutrient dense food on preference and consumption in young and elderly subjects, *Food Quality and Preference*, vol. 8, issue 2, pp 151–156.

Griffin, R, Stauffer, L 1990, Product optimization in central-location testing and subsequent validation and calibration in home-use testing, *Journal of Sensory Studies*, vol. 5, issue 4, pp 231–240.

Gunter, B, Nicholas, D, Huntington, P, Williams, P 2002, Online versus offline research: Implications for evaluating digital media, *Aslib Proceedings*, vol. 54, issue 4, pp 229–239.

Harrar, V, Toepel, U, Murray, M, Spence, C 2011, Food's visually perceived fat content affects discrimination speed in an orthogonal spatial task, *Experimental Brain Research*, vol. 214, issue 3, pp 351–356.

Hennig-Thurau, T, Gwinner, K, Walsh, G, Gremler, D 2004, Electronic word-of mouth via consumer-opinion platforms: What motivates consumers to articulate themselves on the Internet, *Journal of Interactive Marketing*, vol. 18, issue 1, pp 38–52.

Hersleth, M, Ueland, O, Allain, H, Naes, T 2004, Consumer acceptance of cheese, influence of different testing conditions, *Food Quality and Preference*, vol. 16, issue 2, pp 103–110.

Hinkelmann, K, Kempthorne, O 2005, *Design and Analysis of Experiments: Advanced Experimental designs*, vol. 2, John Wiley & Sons, Inc., Hoboken, NJ.

Howe, N, Strauss, W, 2000, *Millennials Rising: The Next Great Generation*, Vintage Books, New York, NY.

Hui, C, Triandis, H 1989, Effects of culture and response format on extreme response style, *Journal of Cross-Cultural Psychology*, vol. 20, issue 3, pp 296–309.

Johnson, D 1997, *Hair and Hair Care*, Marcel Dekker, Inc., New York, NY.

Khare, M, Federer, W 1981, A simple construction procedure for resolvable incomplete block designs for any number of treatments, *Biometrical Journal*, vol. 23, issue 2, pp 121–132.

Kietzmann, J, Hermkens, K, McCarthy, I, Silvestre, B 2011, Social media? Get serious! Understanding the functional building blocks of social media, *Business Horizons*, vol. 54, Issue 3, pp 241–251.

Kumar, A, Lim, H 2008, Age difference in mobile service perceptions: Comparison of generation Y and baby boomers, *Journal of Services Marketing*, vol. 22, issue 7, pp 568–577.

Löfgren, M 2005, Winning at the first and second moments of truth: An exploratory study, *Managing Service Quality: An International Journal*, vol. 15, issue 1, pp 102–115.

Lowengart, O 2012, The effect of branding on consumer choice through blind and non-blind taste tests, *Innovative Marketing*, vol. 8, issue 4, pp 7–18.

Madden, T, Hewett, K, Roth, M 2000, Managing images in different cultures: A cross-national study of color meanings and preferences, *Journal of International Marketing*, vol. 8, issue 4, pp 90–107.

Michel, C, Velasco, C, Spence, C 2015, Cutlery matters: Heavy cutlery enhances diner's enjoyment of the food served in a realistic dining environment, *Flavour*, vol. 4, p 26.

Milotic, D 2003, The impact of fragrance on consumer choice, *Journal of Consumer Behaviour*, vol. 3, issue 2, pp 179–191.

Moran, R, Abramson, N, Moran, S 2014, *Managing Cultural Differences*, Routledge, London and New York.

Piqueras-Fiszman, B, Spence, C 2012, The influence of the color of the cup on consumers' perception of a hot beverage, *Journal of Sensory Studies*, vol. 27, issue 5, pp 324–331.

Piqueras-Fiszman, B, Alcaide, J, Roura, E, Spence, C 2012, Is it the plate or is it the food? Assessing the influence of the color (black or white) and shape of the plate on the perception of the food placed on it, *Food Quality and Preference*, vol. 24, issue 1, pp 205–208.

Piqueras-Fiszman, B, Spence, C 2015, Sensory expectations based on product-extrinsic food cues: An interdisciplinary review of the empirical evidence and theoretical accounts, *Food Quality and Preference*, vol. 40, part A, pp 165–179.

Pocock, S 2010, *Clinical Trials: A Practical Approach*, John Wiley and Son, Ltd, Chichester.

Porcherot, C, Delplanque, S, Raviot-Derrien, S, Le Calvé, B, Chrea, C, Gaudreau, N, Cayeux, I 2010, How do you feel when you smell this? Optimization of the verbal measurement of odor-elicited emotions, *Food Quality and Preference*, vol. 21, issue 8, pp 938–947.

Raghubir, P, Tyebjee, T, Lin, Y 2008, The sense and nonsense of consumer product testing: How to identify whether consumers are blindly loyal? *Foundations and Trends in Marketing*, vol. 3, issue 3, pp 127–176.

Romero, D, Galuba W, Asur, S, Huberman, B 2011, Influence and passivity in Social Media. In: Gunopulos, D, Hofmann, T, Malerba, D, Vazirgiannis, M (eds) *Machine Learning and Knowledge Discovery in Databases*, vol. 6913 Lectures Notes in Computer Science, Springer, Berlin and Heidelberg, pp 18–33.

Safko, L 2010, *The Social Media Bible: Tactics, Tools, and Strategies for Business Success*, John Wiley and Sons, Inc., Hoboken, NJ.

Shankar, M, Levitan, C, Prescott, J, Spence, C 2009, The influence of color and label information on flavor perception, *Chemosensory Perception*, vol. 2, issue 2, pp 53–58.

Si, S, Cullen, J 1998, Response categories and potential cultural bias: Effects of an explicit middle point in cross-cultural surveys, *International Journal of Organizational Analysis*, vol. 6, issue 3, pp 218–230.

Spence, C 2015, Multisensory flavor perception, *Cell*, vol. 161, issue 1, pp 24–35.

Spence, C, Wan, X 2015, Beverage perception and consumption: The influence of the container on the perception of the contents, *Food Quality and Preference*, vol. 39, pp 131–140.

Stening, B, Everett, J 2010, Responses styles in a cross-cultural managerial study, *The Journal of Social Psychology*, vol. 122, issue 2, pp 151–156.

Stone, M, Alisson, B, Foss, B 2004, *Consumer Insight: How to Use Data and Market Research to Get Closer to Your Customer*, Kogan Page Limited, London and Sterling, VA.

Wakeling, I, MacFie, J 1995, Designing consumer trials balanced for first and higher orders of carry-over effect when only a subset of k samples from T may be tested, *Food Quality and Preference*, vol. 6, issue 4, pp 299–308.

Walsh, V 1984, Invention and innovation in the chemical industry: Demand-pull or discovery-push, *Research Policy*, vol. 13, issue 4, pp 211–234.

Wan, X, Woods, A, Seoul, KH, Butcher, N, Spence, C 2015, When the shape of the glass influences the flavour associated with a coloured beverage: Evidence from consumers in three countries, *Food Quality and Preference*, vol. 30, pp 109–116.

Youn, S, Na, J, Choi, S, Huh, C, Park, K 2005, Regional and seasonal variations in facial sebum secretions: A proposal for the definition of combination skin type, *Skin Research and Technology*, vol. 11, issue 3, pp 189–195.

Yu, JH, Albaum, G, Swenson, MJ, 2003, Is a central tendency error inherent in the use of semantic differential scales in different cultures? *International Journal of Market Research*, vol. 45, Issue 2, pp 213–231.

Zampini, M, Sanabria, D, Phillips, N, Spence, C 2007, The multisensory perception of flavour: Assessing the influence of color cues on flavor discrimination responses, *Food Quality and Preferences*, vol. 18, issue 7, pp 975–984.

2

Sensory Profile of a Product: Mapping Internal Sensory Properties

2.1 Origins of Sensory Evaluation

The foundations of sensory evaluation, as is extensively used today, was pioneered in the 1950s and then fully established in academia and industry in the 1960s (Amerine et al. 1965). Its applications were first implemented in many food industries and then expanded to other product categories. These methods have grown in parallel with the development of associated statistical methods especially adapted to analyze multidimensional data (O'Mahony 1985) and computing power capable of managing, in an acceptable time, very large data sets. Rapidly, standards were put in place by the community working on this domain (Hootman 1992) for them to be constantly updated.

2.2 Definition of Descriptive Sensory Analysis

Any packaged consumer good developed to be put on the market has complex and multidimensional sensory properties that all together define its quality. To describe these sensory properties, it is now very common to use descriptive sensory evaluation methods. The goal of these methods is to define the product's sensory identity card encompassing a comprehensive list of objective sensory attributes and for each a rating on an intensity scale. This set of descriptors and their value allow comparing it objectively with other products in the same universe. Indeed, each individual sensory property can be compared among products, and therefore, it is possible to identify and quantify the nature of the differences between products (Carpenter et al. 2000; Drake 2007).

Descriptive sensory evaluation is used in the industry to fulfil different objectives:

- **Quality control** of a product sensorial properties by checking compliance with established parameters and design criteria (aka specifications).

Consumer and Sensory Evaluation Techniques: How to Sense Successful Products, First Edition. Cecilia Y. Saint-Denis.
© 2018 John Wiley & Sons Ltd. Published 2018 by John Wiley & Sons Ltd.

Examples: acidity of a coffee batch, sweetness of an ice tea, foaming properties of a shampoo, consistency of a moisturizing cream and odour defaults in packaging materials (Feria Morales 2002; Etaio et al. 2010).

- **Impact assessment of process or recipe modifications** on sensory properties.

    Examples: impact of certain treatments on food sensorial properties, impact of the replacement of a raw material in a cosmetic formula, influence of change in supply source for an ingredient in a food recipe, differences between two manufacturing locations, and, at large, cost-savings measure that may have an impact in changing the characteristic of a product (Moeller et al. 1999; Lin et al. 2002).

- **Explain consumer preferences**: These are certain sensory properties significantly correlated to consumer ratings and therefore potentially impactful on consumer preferences (Sidel & Stone 1993; Van Trijp & Schifferstein 1995).

    The descriptive sensory information is collected from a group of subjects who have been extensively trained to assess each sensory attribute on intensity scales with established standards.

2.3   Existing Descriptive Methods, Advantages and Disadvantages

Murray et al. wrote in 2001 a comprehensive review of the different descriptive sensory evaluation methods that are still widely used in the industry to characterize sensory product properties. Below appears a summary of these main traditional methods.

2.3.1   Quantitative Descriptive Analysis (QDA)

This method was one of the first to be established and has become the most widely used today (Gacula 2008). It is fully covered in most manuals (few analysis in various field: Shamaila et al. 1992; Chapman et al. 2001; Cadot et al. 2010). Also usually called *Sensory Profile*, it is based on individual ratings collected among trained experts for a comprehensive list of sensory descriptors.

2.3.1.1   Main Characteristics of QDA

- **Environmental control**: Many parameters must be strictly controlled when conducting sensory evaluation to ensure measuring true product differences. These include test room design and colour (typically 50/50 grey), lighting (typically daylight), temperature, humidity, noise, air circulation and odour control, equipment that is used. Also, samples that are presented for evaluation are blind (anonymous and coded) and controlled in their characteristics such as temperature for food products. For a complete understanding of these constraints, see Meilgaard et al. (2007, p.25 and following).

- **Subjects** (also called panellists) are extensively trained experts for a specific product category (using references that belong to the category).
- **Descriptors** (also called attributes) cover comprehensively all the relevant sensory dimensions of the product category (visual, odour, taste, tactile, sound) and are established based on a consensual group work with products in that category (Figure 2.1).

It is commonly established that the list of descriptors must comply with the following criteria: be **objective** (a sensory profile does not include hedonic terms such has *disgusting, pleasant*), be **relevant** to the category (we would not include *sour* to describe texture of a cosmetic cream, for example), be **quantifiable** on a scale, **precise** in their definition, **non-redundant** and **comprehensive** (the list must cover all dimensions with a minimum number of descriptors having no overlap to ensure efficiency and avoid confusion), **discriminant** (descriptors must allow to show differences between products) (translated from Barthelemy et al. 1998, p.151).

Figure 2.1 Sensory descriptors word cloud.

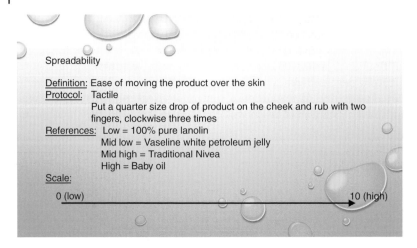

Figure 2.2 Example of sensory descriptor definition.

Methods to establish the final list of descriptors that will constitute the sensory profile have been often described. There are many approaches; some traditionally imply statistical sorting and reduction of initially created large lists of descriptors (Byrne et al. 2001; Carbonell et al. 2007). In practice, lists are often defined in a very pragmatic way. Panellists gather and brainstorm with a large number of products from the category, generate attributes individually and as a group and then consensually reduce the list by selecting and regrouping terms, establishing definitions and references as shown in Figure 2.2 (translated from Urdapilleta et al. 2001, p.93). Today, most product categories benefit from publicly available well-defined lists that can be used and tweaked as needed (Drake & Civille 2003; Lawless & Civille 2013).

Regarding the references, at the time of choice, it is very important to consider references that will be controllable, stable and long-lasting. Very often, when brainstorming with multiple products from the market, some competitors, at times very 'niche' products, may present particularly interesting sensorial characteristics to illustrate certain attributes. Before establishing those references for training, it is necessary to assess how risky that may be. Indeed, at any point in time they may be reformulated (in which case they may lose their reference properties without it being necessarily obvious as the brand may decide not to communicate on the changes) or they may be withdrawn from the market. In some cases, panel leaders may decide to purchase large quantities of those ideal references in the hope of a better option cropping up. This is always a possibility if shelf life is compatible with that.

- **Linear intensity scales,** usually non-structured, are now less time-consuming to compute with existing software. Back in the olden days, panel leaders used to measure values with rulers! Nowadays, they are used to collect evaluations once individuals in the panel are trained (Figure 2.2). The appendix of this chapter shows an example of full sensory questionnaire with a structured scale.

- **Performance** of each subject in the panel, and of the group above all, are regularly checked out with methods now exhaustively described in the latest editions of reference manuals (Meilgaard et al. 2007; Gacula 2008; Lawless and Heymann 2010; Stone et al. 2012) and are included in most sensory evaluation software packages.[1] Methods commonly used encompass simple graphical techniques combined with basics statistics (standard deviations of the measures) and analysis of variance – ANOVA[2] (Lundahl & McDaniel 1988; Naes & Langsrud 1998; Latreille et al. 2006). Since these methods were developed, teams working on this domain have constantly been fine-tuning approaches to assess panels' performance (e.g. Banfield & Harries 1975; Bi 2003).

Different parameters are to be carefully monitored for each attribute:

- **Repeatability** for individuals and for the group: Ability to give the same or significantly close ratings when given the same product blindly, usually within a short period, typically during the same session (Naes & Solheim 1991; Tomic et al. 2007).
- **Reproducibility** for individuals and for the group: Ability to give the same or significantly close ratings when given the same product over a longer period, usually during separated sessions, within days, weeks or months. Sometimes repeatability is also understood as variability between panels or laboratories (Rossi 2001; Tomic et al. 2010).
- **Discrimination**: Ability to significantly differentiate products on the basis of their intensity ratings (Vannier et al. 1999).
- **Ability to use the scale** in its totality: Some panellists are often less inclined to use scale extremes than others (King et al. 2001).

---

1 Most commonly used are:
   SIMS: www.sims2000.com
   Compusense: www.compusense.com
   Fizz: www.biosystemes.com
   PanelCheck: www.panelcheck.com (specifically designed for panel performance checking).
2 The principle is that responses that are obtained Y_{ijk} (individual ratings per product and per descriptors) are dependent on different factors: variability of products i, subjects j, sessions k and interaction among these individual factors. Below is the model not encompassing interactions, ε being the model residual:

$$Y_{ijk} = \mu + \alpha i + \beta j + \delta k + \varepsilon_{ijk}$$

The ANOVA results will show the most determining factors. In the case of sensory profiles, we expect the product effect i to be significant (indicating a significant discrimination), and subject j or session k effects to be nonsignificant (indicating good repeatability or reproducibility). A significant *subject X product* interaction is usually an indicator of poor consensus as it means that each subject has their own way of assessing each product.

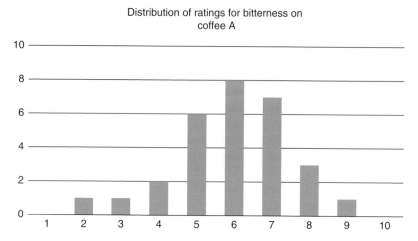

Figure 2.3 Example of an acceptable distribution of sensory ratings.

- **Consensus** in positioning products relatively to one another within the group (also often called **homogeneity**): If a product has consistently higher ratings for a given attribute for most individuals in the group, a panellist with inverted ratings may be considered as an outlier and may need retraining (Bárcenas et al. 2000; McEwan et al. 2003).
- **Normality** of distribution of ratings for each product and each attribute, in the statistical sense (Geary 1947), to be able to use average values in the analysis. Very often actual collected values do not pass the normality test from a significant standpoint in practice (the typical test used is Kolmogorov–Smirnov test, see Lilliefors 1967). Thus, it is necessary to check that values do not show a multimodal distribution but follow a 'bell' curve as shown in Figure 2.3 (Villanueva et al. 2000). If normality of distribution is not obtained, it is recommended to transform the data into ranks and use non-parametric tests (Conover & Iman 1981).

Individual and group performance is primarily based on the quality of the training. For a comprehensive description of training methods, see Meilgaard et al. (2007, p.141 and following). Depending on the complexity of the product category and the number of descriptors, it is quite common to consider and budget between 2 and 6 months of training before being able to routinely use a panel. Of course, it depends on the size of the panel and the number and length of sessions that can be scheduled per week. Efficacy of the training, and hence performance later, will be impacted by when and how often the sessions are scheduled. It seems intuitive that it is preferable to schedule the sessions in a way that corresponds to when the products are usually consumed. Thus, it is often more suitable to test personal care products in the morning but savory products around midday for instance. Common sense and consumer insight knowledge will always prevail to determine the sessions timing. It is also recommended not to have too lengthy sessions to limit habituation or fatigue.

Some institutions or companies rely on internal panels (students, employees), and others rely on strictly outside recruited volunteer subjects. Internal panels are often considered due to budget constraints. However, many have noted that motivation and assiduity can be lower with internal panels than with recruited volunteers. Indeed, employees may not need additional incentive and will have the advantage of being familiar with the category; therefore, they will be easier to train. However, they may feel it is an obligation and a struggle to balance time versus their usual tasks, unless specific time is assigned. In the case of students, as well as recruited volunteers, the primary motivation will rely on the incentive which will warranty loyalty. Loyalty is an extremely important consideration from the budget standpoint. The last thing we want after spending several months training panellists is for them to resign, as this implies having to train new panellists to ensure an acceptable size group. This can delay the moment the panel can operate and can impact overall performance. Many sensory panel initiatives have failed due to the impossibility to keep the panellists active. It is always recommended, for initial training, to 'overbook' the number of panellists that are included in the beginning to anticipate potential withdrawals. The choice of recruiting internal or external panellists will depend on the budget allocated and the context, keeping in mind that the main goals are loyalty and assiduity/punctuality in the sessions. Starting with an internal panel to test feasibility and relevance of the method, and then externalize, can be a quite a wise and commonly adopted strategy in consumer-good companies.

2.3.1.2 Discussion on Inter-Individual Variability

In theory, trained panellists (*aka* human instruments) are so accurate that they are interchangeable and, to a certain extent, one panellist should be sufficient. However, anybody who has worked in this domain knows human measures encompass a significant variability. Overall, training and performance monitoring are conducted to limit this variability as much as possible. To ensure a more robust measure, panels are usually constituted of 10–30 experts. There is no set optimal number of panellists to be included, defined in the pertinent bibliography. Depending on the domain, it may vary as the goal is to obtain an acceptable performance in the end. The number of panellists is also often dictated by logistics constraints (how many panellists can be recruited and seen on a regular basis). Even though individual measures are scrutinized for their reliability, what matters the most in sensory evaluations is the average group measures. Indeed, in sensory analysis, the panel is the measuring tool. Depending on individual performances and logistics, in some cases, it can be decided to conduct two replications of each product measure, while in other cases, one measure may be quite sufficient. In some domains, measures are less variable. Indeed, it is much less subject to variability assessing taste or texture attributes on different chocolate mousses, than to assess dry hair results linked to different shampoos. In the latter, hair being a variable substrate, its variability will impact subjects' acuity. With some hair products implying more tangible results, like hair gels or sprays, it is usually easier to achieve a more acceptable performance than with products like shampoos or conditioners, whose uniqueness in texture or usage properties often fades when it comes to dry hair characteristics due to hair variability.

2.3.1.3   Discussion on Inter-Panel Variability

Results among two trained panels for the same product category are theoretically similar. Many teams have compared panels from different countries on the same products and have found that to be the case (Aparicio and Morales 1995; Pagès & Husson 2001; Lê et al. 2008). However, in practice, results are not always fully transposable (McEwan 1999). This is often linked to the degree of training or expertise of the panellists (Zamora & Guirao 2004; Bitnes et al. 2007) or to the necessity to agree upon a vocabulary to overcome cultural barriers (Hunter & McEwan 1998). Also, reproducing a trained panel in a different context will necessarily induce variability in the measuring tool as explained in a very interesting way by McEwan et al. (2002). Thus, for some product categories, climate may have an impact on sensory properties that the products exhibit. For instance, sensory evaluation is commonly used to assess sensory properties of cosmetic products (Parente et al. 2005). However, it may not be the same to assess a face moisturizing cream in a very dry and cold region compared to a tropical zone, even if the panel is strictly constituted of subjects having the same skin type in both locations and may have been similarly trained. Culture and habits have an impact on acuity to certain sensory dimensions and also on expectations. An interesting example is how consumers from different regions may not have the same sensitivity to boar taint (Font-i-Furnols 2012). In some cultural environments, it can be perceived as unacceptable, whereas in others, it seems normal. Furthermore, some individuals have anosmia to the responsible components due to genetics or to the environment in which they grew up. In such cases, where sensitivity to specific sensorial properties varies, it is very difficult to have comparable measuring tools and is often preferable to specialize panels in the most appropriate region to further exploit results. Hence, food habits will dictate where it is most appropriate to develop a panel for specific products. That is also true in other fields: a panel to assess skin lightening products for instance should be developed in an Asian country where the market is located and a panel to assess hair leave-in treatments should be implemented in South America where it is more prevalent.

Hair products are one of the most challenging to assess via sensory evaluation methods with an acceptable variability (Bouillon & Wilkinson 2005), even more

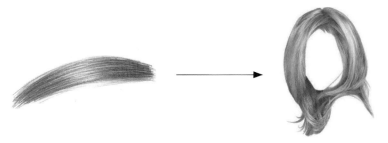

Figure 2.4 From the evaluation on a hair swatch to reality.

when evaluations are compared in different locations. Different industries have experimented on different scenarios to conduct QDA on hair products (Figure 2.4):

- Some attributes can be assessed on **hair swatches** (for instance, all attributes that qualify the interaction of the product with the hair, visual aspects once the product is applied or some tactile dimensions once it is dry, like *smoothness* or *detangling* properties). However, not all relevant attributes can be assessed on hair tresses and often results are quite far from reality (detangling a hair swatch may not be fully representative of detangling an entire head). Some important hair attributes like *curliness* or *volume* are even impossible to asses on hair swatches. In addition, hair inconsistency will necessarily introduce variability in results.
- Another scenario is to have a **'self-applying' trained panel**, in the same way as is done for food: panellists come, apply and evaluate sensory attributes. Panellists need to be recruited based on very precise hair properties and habits to reduce variability linked to those parameters. This approach presents several limitations. First, panellists will never be completely identical and interchangeable in their hair characteristics. Training is aimed at anchoring the references within each panellist's universe and thresholds. However, variability of results will always be relatively high in comparison to other domains where QDA is used. In practice, there are also many logistical constraints, since, in most cases, one person can only evaluate one product at a time. Timing of the sessions during the day (which can be lengthy) must be planned per panellist's habits to facilitate recruitment and assiduity (are they used to wash their hair early in the morning or late in the evening?). Some products require more than one time point assessment, but finding people who accept assessments during multiple day points can be an unsurmountable challenge (either they need to come back to the sensory laboratory or they must do it from work or home, which adds to the uncertainty of what is measured). In some countries, consumers may have routines that

involve multiple products being layered. Finding people who accept to just use one product (a shampoo for instance) and not their typical routine can be a major difficulty, even with a significant incentive. Lastly, in the case of a self-applying trained panel, it is easy to understand that having two comparable panels in two different world locations is extremely difficult to achieve. All this shows how certain product categories having very comparable panels in different locations may be impossible, even undesirable as it may not make sense. Thus, for products that are used or consumed in different regions of the world but in different ways, building different local expert panels is a much more relevant approach, with protocols that will match habits and reality more closely. This will ultimately lead to results that are potentially more predictable of consumer liking.

- The last approach that many cosmetic industries use is to function with **professional panels**, made up of trained cosmetologists who evaluate products applied on volunteers. Logistically, this is easier to implement from timing and budget standpoint although it requires staffed cosmetologists. However, the fact that the evaluator is not the person applying it, will necessarily result in an additional variability factor. Furthermore, if the same volunteer/product combination is evaluated successively by different experts, one must be very cautious in checking that the state of what is being evaluated will not evolve dramatically.

2.3.1.4 Variants to QDA

Quantitative flavour or texture profiles are very similar methods to QDA with the particularity that they exclusively focus on just one sensorial dimension – flavour or texture. Based on the same principles as general QDA, they were also developed in the early 1950–1960s (Caul 1957; Brandt et al. 1963). Very often, one of the objectives of these profiles is to establish correlations between sensorial and instrumental measurements to determine what instrumental parameters are relevant to predict sensorial properties. Many have developed specific flavour or texture profiles applied to their product areas. Below is a non-comprehensive list of interesting examples that can be found in the relevant literature:

- Flavour profile: Narasimhan et al. (1992), Stampanoni (1994), Maetzu et al. (2001), Bárcenas et al. (2001), Lindinger et al. (2008).
- Texture profile: Meullenet et al. (1998), Caine et al. (2003), Philippe et al. (2004).

2.3.1.5 Typical Representations

The most common way to represent sensory profile results is linear sensory profiles (Figure 2.5) or spider diagrams/radar charts (Figure 2.6). Criteria of choice between one or the other is linked to available software or personal preferences. Those are commonly used when the number of products is low.

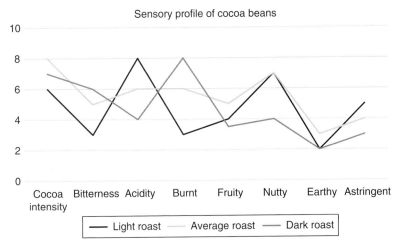

Figure 2.5 Sensory profile of cocoa beans – linear representation.

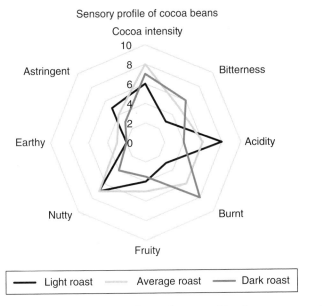

Figure 2.6 Sensory profile of cocoa beans – spider diagram representation.

The habit has always been to tie the points even though the different sensory attributes are independent.

In addition, ANOVA is used to show significant product effects (overall), as well as possible judge and Judge × product interaction effects (Section 2.3.1.1).

Multiple comparison tests (Westfall et al. 2011) need to be run to point out significant differences between paired products. Indeed, ANOVA allows to determine whether there is a significant product effect overall for each attribute, but paired-sample comparisons need to be done to understand which products are different from each other for each attribute. Many softwares can be used to run multiple comparison analysis. In sensory evaluation, SPSS[3] is a commonly used one. Some examples of outputs from an ANOVA followed by a multiple comparison analysis on sensory evaluation data and how to interpret them are given next.

Table 2.1 shows sweetness intensity values for three products A, B and C. All products were evaluated by five subjects. Product A has a mean value of 4.2 for that attribute, product B has a mean value of 3.6 and product C has a mean value of 2.2. The standard deviations of their sweetness intensities are 1.30, 0.89 and 0.8, respectively.

ANOVA is used to determine if the means of the different products in this case are statistically different. In the example shown in Table 2.2, variability of the sweetness intensity values within each group (products A, B or C) is significantly lower than the variability between groups. The significance (Sig. value or p-value) is the indicator of whether there are significant differences between products or not. If Sig. is greater than 0.05, it is common practice to say there are no statistical differences between products. If Sig. is less than or equal to 0.05, conclusion is that there are statistically significant differences between products.

However, the Sig. value does not allow us to determine which product means are different from each other. It only says that overall there are significant differences among the products. For example, if among eight products, one is completely different from the seven others and the seven others are quite similar; Sig. value will still show up as significant. Hence, the need to dig deeper and understand what differences exist between paired products by conducting multiple comparisons, often called 'Post-hoc tests'. In Figure 2.5, for the attribute nutty, the ANOVA results will most likely show an overall significant product effect. But it is the dark roast sample that may be driving the effect, whereas the two other products are almost identical for that attribute.

For all-pairwise comparison in a sensory evaluation context, the usual recommended method is Tuckey, as it is more conservative when testing large number of means compared to others like Bonferroni or Least Significant Differences (LSD). To better understand the different tests and which ones are recommended in which conditions, one can refer to Hsu (1996) or Westfall et al. (2011). Most software, like SAS or SPSS, allow to easily implement the different methods and give recommendations based on the dataset. Often, one can try different algorithms and notice either very similar results or some being more conservative than others, which may give elements to validate which differences are truly significant and which may be interpreted as trends.

---

3 https://www.ibm.com/us-en/marketplace/spss-statistics

Table 2.1 Mean comparison, example of descriptive statistics.

Products	N	Mean	Std. deviation	Std. error	95% Confidence interval for mean		Min	Max
					Lower bound	Upper bound		
A	5	4.2000	1.3038	0.5831	2.5811	5.8189	3.00	6.00
B	5	3.6000	0.8944	0.4000	2.4894	4.7106	3.00	5.00
C	5	2.2000	0.8367	0.3742	1.1611	3.2389	1.00	3.00
Total	15	3.3333	1.2910	0.3333	2.6184	4.0483	1.00	6.00

Table 2.2 Mean comparison, example of ANOVA results.

	Sum of squares	df	Mean square	F	Sig.
Between groups	10.533	2	5.267	4.938	0.027
Within groups	12.800	12	1.067		
Total	23.333	14			

Table 2.3 Mean comparison, example of multiple comparison analysis.

					95% Confidence interval for mean	
(I)	(J)	Mean difference (I − J)	Std. error	Sig.	Lower bound	Upper bound
A	B	0.6000	0.6532	0.639	−1.1426	2.3426
	C	2.0000[a]	0.6532	0.025	0.2574	3.7426
B	A	−0.6000	0.6532	0.639	−2.3426	1.1426
	C	1.4000	0.6532	0.123	−0.3426	3.1426
C	A	−2.0000[a]	0.6532	0.025	−3.7426	−0.2574
	B	−1.4000	0.6532	0.123	−3.1426	−0.3426

[a] Mean difference is significant at the 0.05 level.

Table 2.3 shows an example of multiple comparison analysis (Tuckey) for the sweet intensity between products A, B and C. Here the Sig. value shows whether the two products that are being compared are significantly different. If Sig. value is less than or equal to 0.05, the two products are significantly different for this attribute. In the example shown, it is the case for products A and C. If Sig. value is greater than 0.05 there is no statistically significant difference between the two products.

When the number of products augments, linear profiles or spider charts may become hard to read and interpret. In those cases, principal component analysis (PCA) (read Borgognone et al. 2001 on how to define parameters for PCA and whether it is more appropriate to use covariance or correlation matrix depending on the data set) is commonly used to represent the multidimensional data in a 2- or 3-dimensional map. There are different schools of thoughts on whether attributes that do not show a significant difference between products should be included in the PCA. In theory, they should not, as the risk is to introduce noise. However, in most cases, introducing them does not impact the results globally and the representation as they end up being projected close to the centre of the map with low weights in the overall system.

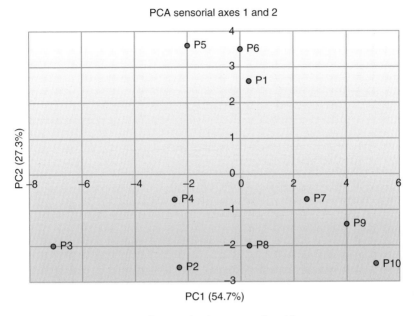

Figure 2.7 Sensory map: products projection on axes 1 and 2.

In Figures 2.7 and 2.8, the two maps are to be interpreted in parallel. Figure 2.7 shows the projection of the products in the sensory space (coffees in this map). Figure 2.8 shows the projection of the sensory attributes. Overall, the data set is relatively well represented in these two dimensions as they summarize 82% of the total variance (sum of 54.7% with axis 1 and 27.3% of axis 2). This means that the representation contains 82% of the information contained in the original data set. Sometimes it is necessary to look at axis 3 or higher to understand positioning for certain attributes. Products are placed in the same space area than the sensory attributes for which they have high ratings and opposed to the sensory attributes for which they have low ratings. Therefore, the PCA shows a relative placement of these products based on their ratings for this set of sensory attributes. In our example, axis 1 on the right, is characterized by strong acidity *aci* and remanence *rem* but low nutty *nut* and floral *flo* scores. Axis 2 on the top, represents strong bitterness *bit* and body *bod* but low burnt *bur* and earthy *ear* values. Attributes that are close in space are correlated (which could show the attributes are intrinsically linked, in which case the list could be reduced, but some correlations can be just *de facto* for a particular dataset). In our example, acidity and remanence tend to have correlated ratings. However, this does not mean that both attributes are not valuable. Indeed, there could be products for which acidity and remanence are independently rated in the coffee universe. In the same way, in our example, acidity and nutty are anti-correlated.

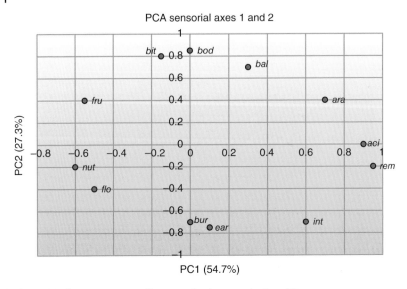

Figure 2.8 Sensory map: attributes projection on axes 1 and 2.

However, these two attributes are not necessarily always anti-nomic. Attributes that are projected on perpendicular directions are independent (example of acidity and bitterness). We can summarize how to read a sensory map like this: if we take coffee P3, it is characterized by strong values for floral and nutty, relatively low values for acidity, remanence and arabica *ara*, slightly burnt *bur* and earthy notes, low balance *bal*. It is distinct from the other coffees. It is important to underline that PCA representations of QDA sensory data are relative. This means that product positioning is always relative to one another. Thus, within this particular set, a product positioned as the most acid may not necessarily be the most acid in the world of coffees. However, that information is given somehow by the absolute rating, although it must always be nuanced by the fact that panellists always assess keeping in mind the universe they know. If a product comes up and goes beyond the established references for a particular attribute, then the references will have to be ultimately shifted.

To interpret PCA representations, it is also necessary to look at all additional outputs. These usually include:

- Correlation or covariance matrix for attributes which shows which attributes are correlated or anti-correlated to each other
- Product and attribute coordinates in their respective representations
- Percentage of variance explained by each component (Table 2.4)

It is common practice to look primarily at the axis showing an eigenvalue higher than or equal to 1. Eigenvalues are mathematical parameters associated with each principal component that tell how much variation in the data set

Table 2.4 PCA, example of total variance explained.

Component	Eigenvalues		
	Total	% of variance	Cumulative (%)
1	6.249	52.076	52.076
2	1.229	10.246	62.322
3	0.719	5.992	68.313
4	0.613	5.109	73.423
5	0.561	4.676	78.099
...	...	...	...

they explain. Hence, an eigenvalue superior or equal to 1 means that the axis brings at least as much information as one of the original variables (attributes) included in the analysis. An axis which eigenvalue is lower than 1 may only represent noise.

More recently, some teams have developed new representation approaches based on methods such as multidimensional scaling (Cox & Cox 2001) or on intelligent techniques such as fuzzy logic, neural networks and data aggregation. These seem interesting to understand the sometimes uncertain or imprecise sensory data (Ruan & Zeng 2004). Indeed, neural network–type modelizations are attractive as they can adapt to even complex phenomenon, often non-linear and with multiple interactions. However, once the structure of the network is defined and the learning process completed, a large dataset not used to build the network needs to be used to validate the model. In the end, classic modelization methods are often sufficient and less risky (Section 5.4).

2.3.2 Free-Choice Profile

This method was introduced by Williams and Langron (1984). Its main difference with QDA lies in the fact that each subject develops its own descriptors' list, with their own definition and comprehension. Thereby, its advantage is to overcome inter-individual differences in the way each person understands and perceives different sensorial dimensions. Nonetheless, besides the difference in vocabulary used, all subjects follow the same evaluation protocol. Once all data is collected, generalized Procrustes statistics[4] is used to rationalize the

---

4 Many software such as SAS (www.sas.com), XLstats (www.xlstat.com), FactoMineR (www.factominer.free.fr) (an R package, which has the advantage of being free) and Matlab (www.mathworks.com/products/matlab.html) can be used to perform this type of analysis.

spatial configurations from individual profiles. This method relies on a series of rotations, translations and homotheties to understand how products are positioned for the overall group, by overpassing the differences in vocabulary and scale usage. Williams and Arnold (1985) showed that this method can give very similar product positioning compared to conventional sensory profiles and save time needed to obtain a consensual list of descriptors and train the group according to that list. This method has been since then extensively used in multiple product categories. The main limit to this approach is that the meaning of each sensorial attribute, ultimately shown in the representation, will be based on the analyst's interpretation. Furthermore, those profiles that are obtained can be strongly linked to the group of subjects that participated in the study. In some cases, this approach is used to generate and sort sensorial attributes that are then used in a more classical QDA approach.

2.3.3 Flash Profile

The Flash profile method was established by Pr J.M. Sieffermann's team in the early 2000s (Dairou & Sieffermann 2002; Delarue & Sieffermann 2004) and since then, it has been commonly used as another faster alternative to conventional QDA (Lassoued et al. 2008; Albert et al. 2011). This method derives from the free-choice profile. However, in the flash profile, all samples are presented simultaneously for subjects to rank according to those descriptors selected as the most relevant to show differences among the products.

2.3.4 Spectrum

The philosophy of this method established by Munoz and Civille (1998) is based on a universal tool that can be used to describe all types of products. A bank of descriptors is available with universal definitions and references (usually called 'external' references as opposed to 'internal' references used in QDA, meaning references that do not belong exclusively to a specific category). Spectrum method offers a lot of freedom to the panel leader. It only provides a descriptor database for a broad range of universal sensory dimensions illustrated with absolute references (belonging to a wide range of product categories) to exemplify a scale going from 0 to 15 (usually 3–5 references per attribute). This method is interesting as it allows collecting absolute values for sensory dimensions on a given product. Hence, its strength is to overcome the relativity limit of QDA. However, it requires intensive and lengthy training to ensure accuracy and reliable measures. Since scales are absolute across all product categories, the risk is to obtain low discrimination for certain attributes when working on a very specific product category.

2.3.5   Time Intensity

The time intensity T-I method, first introduced by Powers and Pangborn (1978), focuses on the dynamic evolution of perceptions over time. In this method, panellists are trained to rate continuously, over a time period, the intensity of sensory attributes. Hence, it is particularly relevant when the sensory sensations evolve over time (Harrison & Bernhard 1984; Lee & Pangborn, 1986). It has grown from paper–pen collection, in its origins, to computerized data collection[5] (Guinard et al. 1985; Lee III 1985), which dramatically eased and boosted its expansion (Cliff & Heymann 1993). Overall, it was developed to be particularly suited to study the dynamics of flavour release in the mouth or of the chewing mechanisms, but it can be technically applied to any area where an overall intensity may not represent what happens over time. For example, it can be very interesting to understand the evolution, in the mouth, of peppermint in chewing gums or tenderness in meat (Naish et al. 1995; Duizer et al. 1997; Dijksterhuis & Piggott 2000). It is also a suitable and powerful tool to assess various artificial and natural sweeteners and sugar alternatives in comparison to the time-intensity profile of 'real' sugar (Medeiros de Melo et al. 2007).

Conducting a time-intensity study should only be considered when it is truly relevant. Indeed, it encompasses multiple constraints. First, it requires recruiting panellists that have a real acuity for the sensations that are to be measured. Panellists need to be extensively trained in traditional sensory methods and in addition to the specificities of T-I. They must simultaneously manage ambivalence between intensity and time and learn to use the computerized interface (usually a mouse and a screen). Furthermore, the analysis of the data that is generated can be very complex as each evaluation will result in an individual curve and potentially some abnormalities to integrate or smooth out. The most common approach to analyze the data has been to extract relevant parameters such as time to maximum intensity, intensity at the maximum, total duration, and so forth (Thorngate III & Noble 1995).

Nowadays, T-I methods are deemed to be commonly used in the food industry when *ad hoc* personnel and material are available. Of course, it makes sense to have a trained T-I panel when its sufficient use is going to be made. In other industries, such as the cosmetics industry, some areas require temporal considerations. For example, the wearing of a nail polish (over a few days or weeks for some technologies) or a lipstick (along the day), the hold of hair curls or hair volume all day long, the skin dryness or hydration over time. In many cases, from a logistics standpoint, it is quite hard or even impossible to

---

5 Most sensory packages now allow to draw up questionnaires, collect and analyze data on continuous time intensity. Most have the multiattribute capabilities (Sims2000, Compusense, fizz).

implement T-I methods. However, QDA method can be adapted by the addition of multiple time points of the same attributes and the analysis will consist in understanding how the ratings evolve through some graphical visualizations and statistics to assess whether an increase or a drop in values are significant. Adding certain time points during the day or weeks after the initial T_0 evaluation will very often introduce variability due to the different non-controlled context. Indeed, in most cases, panellists will assess T_0 in the sensory lab rigorous environment and following time points from a remote location (work, home). It is important to make the panellists fully aware of what is expected to minimize measure bias.

2.3.6 Comparative Advantages and Limits in Each Method

Table 2.5 gives a general overview of what the different previously described methods can bring about and what the limits of each one are.

Teams from industry and academia regularly expand on these core methods to experiment and customize approaches that are easier and quicker to implement while still giving meaningful and reliable results (Cartier et al. 2006; Albert et al. 2011). Varela and Ares (2008) provide an interesting review of a few new sensory profiling techniques and how they compare with conventional QDA or free choice profiling. An interesting and rapid approach that has been implemented in some industries is the Napping® technique (Perrin et al. 2008; Pagès et al. 2010). Incidentally, the name of the method comes from the French term 'nappe' which means 'tablecloth', as the technique consists in having the subjects allocate, sort and group the products on a tablecloth space based on their similarities or differences to ultimately obtain a relative spatial distribution of the products and then reveal explicit groups with qualitative characteristics that can be added. Coordinates of the products and groups they belong to are encompassed in a global statistical analysis. The method has revealed to be efficient in multiple cases and is less time consuming than classical profiling. Also for some product categories, like high alcohol products, it can be interesting when classical sensory methods may be challenging due to sensory fatigue (Louw et al. 2013). However, individual semantic interpretations may be complex to analyze (Dehlholm et al. 2012).

Because the most limiting factor of sensory descriptive methods is the time required to train panels and the risk of losing panellists afterwards, some teams have even spared completely the training time by using regular non-trained consumers to perform QDA. In some cases, results show high similarities (Worch et al. 2010). However, this approach can be risky as accuracy is often not the same with non-trained panels which are still highly recommended when the objective is to obtain precise and objective sensory profiles and descriptions (Bárcenas et al. 2004).

To stay abreast of constant innovations in the sensory field, several sources of information can be found, besides specialized literature where industry and

Table 2.5 Summary of different descriptive sensory analysis methods.

Method	Advantages	Limits	Perspectives
QDA	Precise sensory ID card	Time/budget required to establish descriptors and train the panel	Additional time points can be considered (which may often be measured with added variability)
	Well-established statistical analysis methods	Need to monitor performance and retrain	
		Relative positioning of products	
		No temporal consideration	
Free-choice profile	Much quicker to implement than QDA as it is unnecessary to establish a common lexicon	Sensory profiles are valid only for the group of subjects that performed the measures	It is a good cost-efficient qualitative approach. Many may use it as a preliminary approach to QDA
	Can be performed on non-trained panellists	Analysis of sensory attributes signification can be complex and may be biased by analyst's interpretation	
		The final list of descriptors may not be comprehensive	
Flash profile	Derives from free-choice but faster to implement as products are presented simultaneously	In addition to free-choice profile limits:	Also, a cost-efficient alternative approach. Often used as a preliminary approach to QDA
	It usually enhances discrimination as panellists will evaluate by comparison and ranking	• the number of products that can be evaluated is limited (due to saturation and fatigue) • Results are relative to the product set presented	
Spectrum	External references usually facilitate training and agreement among panellists	Training is usually longer than for QDA	In practice, many add internal references to expand scale usage and enhance discrimination
	Standardization and universalization facilitates expanding and generating measuring tools on multiple categories	Universal scales present the risk of a lower sensitivity to smaller differences and can lead to lower discrimination among products within a specific category	
Time-intensity	Allows analyses of the evolution of a sensorial dimension over time	Time/budget required to implement (training, infrastructure and tools)	Can be complementary to classical QDA for specific attributes
		Complex data analysis	
		Limited number of attributes can be included	

academia teams regularly publish their research. The Society of Sensory Professionals (http://www.sensorysociety.org) is an affluent organisation worth being in touch with. The bi-annual Pangborn symposium (http://www.pangbornsymposium.com) is also a very inspiring way to connect with worldwide sensory and consumer science experts.

2.3.7 Cost Considerations

Starting a sensory evaluation activity often implies having to consider significant investments in time and money. Whether it is developed internally or externally, sensory panels are specialized measuring tools and even if the logistics is handled outside, they are often developed in as 'proprietary' (which means to be used exclusively for the company for which they were developed).

Here is a breakdown of the various investments one must plan for to set such capabilities.

Material investments:

- Sensory evaluation facility with a multi-purpose meeting room (for training sessions and discussions) and individual booths (for evaluations). A multi-purpose room with a round table (seating 8–12 people) can be alternatively used for sensory evaluation training sessions and focus groups (Section 3.1.3), in which case a two-way mirror is needed. In the same way, individual booths can be alternatively used for sensory evaluation sessions and self-application qualitative studies (Section 3.1.3, consumers come to a CLT to try a product and be interviewed). In the latter, it is necessary to install cameras to be able to monitor what happens in the booths.
- Overall proper ventilation, temperature and humidity control need to be set up.
- Depending on the type of product being tested, it is important to consider elements like a chair and a table or bench, a sink or a shower and eventually water with controlled temperature, flow and softness.
- Cold light (daylight) with a colour temperature of 5600–6200 K is recommended to be put in place in most situations (enables a uniform and glare-free aspect). To mute the impact of product colour, a red light may also be added.
- A data acquisition system is to be incorporated.

Human investments:

- A panel leader is needed to handle training and evaluation sessions. This encompasses scheduling of the sessions and communication with panellists as well as animation of the sessions.
- A data analyst is needed to set protocols, analyze and interpret results.
- Panellists: their number and their time will depend on the type of products and the frequency of tests to be carried out.

It is important to consider that an expert panel will not be operational until the training period is completed and performance obtained is acceptable, which may take 2–6 months. During that time, costs incurred will not translate into concrete product results, which is often a deceptive period to overcome. Therefore, it is better to incorporate upfront in the budget the cost of the training period as well as the cost of possible 'maintenance' periods.

References

Albert, A, Varela, P, Salvador, A, Hough, G, Fiszman, S 2011, Overcoming the issues in the sensory description of hot served food with a complex texture. Application of QDA®, flash profiling and projective mapping using panels with different degrees of training, *Food Quality and Preference*, vol. 22, issue 5, pp 463–473.

Amerine, M, Pangborn, R, Roessler, E 1965, *Principles of Sensory Evaluation of Food*, Academic Press, New York and London.

Aparicio, R, Morales, M 1995, Sensory wheels: a statistical technique for comparing QDA panels: application to virgin olive oil, *Journal of the Science of Food and Agriculture*, vol. 67, issue 2, pp 247–257.

Banfield, G, Harries, J 1975, A technique for comparing judges' performance in sensory tests, *Food Science and Technology*, vol. 10, issue 1, pp 1–10.

Bárcenas, P, Pérez Elortondo, F, Albisu, M 2000, Selection and screening of a descriptive panel for ewes milk cheese sensory profiling, *Journal of Sensory Studies*, vol. 15, issue 1, pp 79–99.

Bárcenas, P, Pérez Elortondo, F, Salmerón, J, Albisu, M 2001, Sensory profile of ewe's milk cheeses, *Food Science and Technology International*, vol. 7, issue 4, pp 347–535.

Bárcenas, P, Pérez Elortondo, F, Albisu, M 2004, Projective mapping in sensory analysis of ewes milk cheeses: a study on consumers and trained panel performance, *Food Research International*, vol. 37, issue 7, pp 723–729.

Barthelemy, J, Clement, J, Danzart, M, Issanchou, S, Köster, E, Mac Leod, P, Nicod, H, Sauvageot, F, Trigler, F, Touraille, C 1998, *Evaluation Sensorielle: Manuel Methodologique*, Tec & Doc Lavoisier, Paris.

Bi, J 2003, Agreement and reliability assessments for performance of sensory descriptive panel, *Journal of Sensory Studies*, vol. 18, issue 1, pp 61–76.

Bitnes, J, Rødbotten, M, Lea, P, Ueland, Ø, Martens, M 2007, Effect of product knowledge on profiling performance comparing various sensory laboratories, *Journal of Sensory Studies*, vol. 22, issue 1, pp 66–80.

Borgognone, M, Bussi, J, Hough, G 2001, Principal component analysis in sensory analysis: covariance or correlation matrix? *Food Quality and Preference*, vol. 12, issue 5–7, pp 323–326.

Bouillon, C, Wilkinson, J 2005, *The Science of Hair Care*, Second Edition, CRC Press, Boca Raton, FL.

Brandt, M, Skinner, E, Coleman, J 1963, Texture profile method, *Journal of Food Science*, vol. 28, issue 4, pp 404–409.

Byrne, D, O'Sullivan, M, Dijksterhuis, G, Bredie, W, Martens, M 2001, Sensory panel consistency during development of a vocabulary warmed-over flavour, *Food Quality and Preference*, vol. 12, issue 3, pp 171–187.

Cadot, Y, Caillé, S, Samsom, A, Barbeau, G, Cheynier, V 2010, Sensory dimension of wine typicality related to a terroir by Quantitative Descriptive Analysis, Just About Right analysis and typicality assessment, *Analytica Chimica Acta*, vol. 660, issue 1–2, pp 53–62.

Caine, W, Aalhus, J, Best, D, Dugan, M, Jeremiah, L 2003, Relationship of texture profile analysis and Warner-Bratzler shear force with sensory characteristics of beef rib steaks, *Meat Science*, vol. 64, issue 4, pp 333–339.

Carbonell, L, Izquierdo, L, Carbonell, I 2007, Sensory analysis of Spanish mandarin juices. Selection of attributes and panel performance, *Food Quality and Preference*, vol. 18, issue 2, pp 329–341.

Carpenter, R, Lyon, D, Hasdell, T 2000, *Guidelines for Sensory Analysis in Food Product Development and Quality Control*, Second Edition, Aspen Publishers, Inc., Gaithersburg, MD.

Cartier, R, Rytz, A, Lecomte, A, Poblete, F, Krystlik, J, Belin, E, Martin, N 2006, Sorting procedure as an alternative to quantitative descriptive analysis to obtain a product sensory map, *Food Quality and Preference*, vol. 17, issues 7–8, pp 562–571.

Caul, J, 1957, The profile method of flavor analysis, *Advances in Food Research*, vol. 7, pp 1–40.

Chapman, K, Lawless, H, Boor, K 2001, Quantitative descriptive analysis and principal component analysis for sensory characterization of ultrapasteurized milk, *Journal of Dairy Science*, vol. 84, issue 1, pp 12–20.

Cliff, M, Heymann, H 1993, Development and use if time-intensity methodology for sensory evaluation: a review, *Food Research International*, vol. 26, issue 5, pp 375–385.

Conover, W, Iman, R 1981, Rank transformations as a bridge between parametric and nonparametric statistics, *The American Statistician*, vol. 35, issue 3, pp 124–129.

Cox, T, Cox, M 2001, *Multidimensional Scaling*, Second Edition, Chapman & Hall/CRC, Boca Raton, FL.

Dairou, V, Sieffermann, J 2002, A comparison of 14 jams characterized by conventional profile and a quick original method, the flash profile, *Journal of Food Science*, vol. 67, issue 2, pp 826–834.

Dehlholm, C, Brockhoff, P, Meinert, L, Aaslyng, M, Bredie, W 2012, Rapid descriptive sensory methods: comparison of free multiple sorting, partial napping, napping, flash profiling and conventional profiling, *Food Quality and Preference*, vol. 26, issue 2, pp 267–277.

Delarue, J, Sieffermann, J 2004, Sensory mapping using flash profile. Comparison with a conventional descriptive method for the evaluation of the flavour of dairy products. *Food Quality and Preference*, vol. 15, issue 4, pp 383–392.

Dijksterhuis, G, Piggott, J 2000, Dynamic methods of sensory analysis, *Trends in Food Science and Technology*, vol. 11, issue 8, pp 284–290.

Drake, M 2007, Invited review: sensory analysis of dairy foods, *Journal of Dairy Science*, vol. 90, issue 11, pp 4925–4937.

Drake, M, Civille, G 2003, Flavor lexicons, *Comprehensive Reviews in Food Science and Food Safety*, vol. 2, issue 1, pp 33–40.

Duizer, L, Bloom, K, Findlay, C 1997, Dual-attribute time-intensity sensory evaluation: a new method for temporal measurements of sensory perception, *Food Quality and Preference*, vol. 8, issue 4, pp 261–269.

Etaio, I, Albisu, M, Ojeda, M, Gil, P, Salmerón, J, Pérez Elortondo, F 2010, Sensory quality control for food certification: a case study on wine. Method development, *Food Control*, vol. 21, issue 4, pp 533–541.

Feria Morales, A 2002, Examining the case of green coffee to illustrate the limitations of grading systems/expert tasters in sensory evaluation for quality control, *Food Quality and Preference*, vol. 13, issue 6, pp 355–367.

Font-i-Furnols, M 2012, Consumer studies on sensory acceptability of boar taint: a review, *Meat Science*, vol. 92, issue 4, pp 319–329.

Gacula, M 2008, *Descriptive Sensory Analysis in Practice*, John Wiley & Sons, Inc., Hoboken, NJ.

Geary, R 1947, Testing for normality, *Biometrika*, vol. 34, issue 3/4, pp 209–242.

Guinard, J, Pangborn, R, Shoemaker, C 1985, Computerized procedure for time-intensity sensory measurements, *Journal of Food Science*, vol. 50, issue 2, pp 543–544.

Harrison, S, Bernhard, R, 1984 Time-intensity sensory characteristics of saccharin, xylitol and galactose and their effect on the sweetness of lactose, *Journal of Food Science*, vol. 49, issue 3, pp 780–786.

Hootman, R 1992, *Manual on Descriptive Analysis Testing for Sensory Evaluation*, ASTM Manual Series MNL 13, ASTM, Philadelphia, PA.

Hsu, J 1996, *Multiple Comparisons, Theory and Methods*, Chapman & Hall/CRC, Boca Raton, London, New York, and Washington, DC.

Hunter, E, McEwan, J 1998, Evaluation of an international ring trial for sensory profiling of hard cheese, *Food Quality and Preference*, vol. 9, issue 5, pp 343–354.

King, M, Hall, J, Cliff, M 2001, A comparison of methods for evaluating the performance of a trained sensory panel, *Journal of Sensory Studies*, vol. 16, issue 6, pp 567–581.

Lassoued, N, Delarue, J, Launay, B, Michon, C 2008, Baked product texture: correlations between instrumental and sensory characterization using Flash profile, *Journal of Cereal Science*, vol. 48, issue 1, pp 133–143.

Latreille, J, Mauger, E, Ambroisine, L, Tenenhaus, M, Vincent, M, Navarro, S, Guinot, C 2006, Measurement of the reliability of sensory panel performances, *Food Quality and Preference*, vol. 17, issue 5, pp 369–375.

Lawless, L, Civille, G 2013, Developing lexicons: a review, *Journal of Sensory Studies*, vol. 28, issue 4, pp 270–281.

Lawless, H, Heymann, H 2010, *Sensory Evaluation of Foods, Principles and Practices*, Second Edition, Springer, New York, Dordrecht, Heidelberg, and London.

Lê, S, Pagès, J, Husson, F 2008, Methodology for comparison of sensory profiles provided by several panels: application to a cross-cultural study, *Food Quality and Preference*, vol. 19, issue 2, pp 179–184.

Lee III, W 1985, Evaluation of time-intensity sensory responses using a personal computer, *Journal of Food Science*, vol. 50, issue 6, pp 1750–1751.

Lee III, W, Pangborn, M 1986, Time-intensity: the temporal aspects of sensory perception, *Food Technology*, vol. 40, issue 11, pp 71–78.

Lilliefors, H 1967, On the Kolmogorov-Smirnov test for normality with mean and variance unknown, *Journal for the American Statistical Association*, vol. 62, issue 318, pp 399–402.

Lin, S, Huff, H, Hsieh, F 2002, Extrusion process parameters, sensory characteristics, and structural properties of a high moisture soy protein meat analog, *Journal of Food Science*, vol. 67, issue 3, pp 1066–1072.

Lindinger, C, Labbe, D, Pollien, P, Rytz, A, Juillerat, M, Yeretzian, C, Blankl, I 2008, When machine tastes coffee: instrumental approach to predict the sensory profile of expresso coffee, *Analytical Chemistry*, vol. 80, issue 5, pp 1574–1581.

Louw, L, Malherbe, S, Naes, T, Lambrechts, M, Van Rensburg, P, Nieuwoudt, H 2013, Validation of two Napping® techniques as rapid sensory screening tools for high alcohol products, *Food Quality and Preference*, vol. 30, issue 2, pp 192–201.

Lundahl, D, McDaniel, M 1988, The panelist effect: fixed or random, *Journal of Sensory Studies*, vol. 3, issue 2, pp 113–121.

Maetzu, L, Sanz, C, Andueza, S, Paz de Peña, M, Bello, J, Cid, C 2001, Characterization of espresso coffee aroma by static headspace GC-MS and sensory flavor profile, *Journal of Agricultural and Food Chemistry*, vol. 49, issue 11, pp 5437–5444.

McEwan, J 1999, Comparison of sensory panels: a ring trial, *Food Quality and Preference*, vol. 10, issue 3, pp 161–171.

McEwan, J, Hunter, E, Van Germert, L, Lea, P 2002, Proficiency testing for sensory profile panels: measuring panel performance, *Food Quality and Preference*, vol. 13, issue 3, pp 181–190.

McEwan, J, Heiniö, R, Hunter, L, Lea, P 2003, Proficiency testing for sensory ranking panels: measuring panel performance, *Food Quality and Preference*, vol. 14, issue 3, pp 247–256.

Medeiros de Melo, L, Bolini, H, Efraim, P 2007, Equisweet milk chocolates with intense sweeteners using time-intensity method, *Journal of Food Quality*, vol. 30, issue 6, pp 1056–1067.

Meilgaard, M, Civille, G, Carr, B 2007, *Sensory Evaluation Techniques*, Fourth Edition, CRC Press, Boca Raton, FL.

Meullenet, J, Lyon, B, Carpenter, J, Lyon, C 1998, Relationship between sensory and instrumental texture profile attributes, *Journal of Sensory Studies*, vol. 13, issue 1, pp 77–93.

Moeller, S, Wulf, D, Meeker, D, Ndife, M, Sundararajan, N, Solomon, M 1999, Impact of hydrodyne process on tenderness, microbial load and sensory characteristics of pork longissimus muscle, *Journal of Animal Science*, vol. 77, issue 8, pp 2119–2123.

Munoz, A, Civille, G 1998, Universal, product and attribute specific scaling and the development of common lexicons in descriptive analysis, *Journal of Sensory Studies*, vol. 13, issue 1, pp 57–75.

Murray, J, Delahunty, C, Baxter, I 2001, Descriptive sensory analysis: past, present, future, *Food Research International*, vol. 34, issue 6, pp 461–471.

Naes, T, Langsrud, Ø 1998, Fixed or random assessors in sensory profiling, *Food Quality and Preference*, vol. 9, issue 3, pp 145–152.

Naes, T, Solheim, R 1991, Detection and Interpretation of variation within and between assessors in sensory profiling, *Journal of Sensory Studies*, vol. 6, issue 3, pp 159–177.

Naish, M, Clifford, M, Birch, G 1995, Sensory astringency of 5-O-caffeoylquinic acid, tannic acid and grape-seed tannin by time-intensity procedure, *Journal of the Science of Food and Agriculture*, vol. 61, issue 1, pp 57–64.

Narasimhan, S, Chand, N, Rajalaksmi, D 1992, Saffron: quality evaluation by sensory profile and gas chromatography, *Journal of Food Quality*, vol. 15, issue 4, pp 303–314.

O'Mahony, M 1985, *Sensory Evaluation of Food, Statistical Methods and Procedures*, Marcel Dekker, Inc., New York.

Pagès, J, Husson, F 2001, Inter-laboratory comparison of sensory profiles: methodology and results, *Food Quality and Preference*, vol. 12, issue 5–7, pp 297–309.

Pagès, J, Cadoret, M, Lê, S 2010, The sorted napping: a new holistic approach in sensory evaluation, *Journal of Sensory Studies*, vol. 25, issue 5, pp 637–658.

Parente, M, Gámbaro, A, Solana, G 2005, Study of sensory properties of emollients used in cosmetics and their correlation with physicochemical properties, *International Journal of Cosmetic Science*, vol. 27, issue 6, pp 354.

Perrin, L, Symoneaux, R, Maître, I, Asselin, C, Jourjon, F, Pagès, J 2008, Comparison of three sensory methods for use with the Napping® procedure: case of ten wines from Loire valley, *Food Quality and Preference*, vol. 19, issue 1, pp 1–11.

Philippe, F, Schacher, L, Adolphe, D, Dacremont, C 2004, Tactile feeling: sensory analysis applied to textile goods, *Textile Research Journal*, vol. 74, issue 12, pp 1066–1072.

Powers, N, Pangborn, R 1978, Paired comparison and time-intensity measurements of the sensory properties of beverages and gelatines containing sucrose or synthetic sweeteners, *Journal of Food Science*, vol.43, issue 1, pp 41–46.

Rossi, F 2001, Assessing sensory panelist performance using repeatability and reproducibility measures, *Food Quality and Preference*, vol. 12, issue 5–7, pp 467–479.

Ruan, D, Zeng, X 2004, *Intelligent Sensory Evaluation: Methodologies and Applications*, Springer, Berlin and New York.

Shamaila, M, Powrie, W, Skura, B 1992, Sensory evaluation of strawberry fruit stored under modified atmosphere packaging (MAP) by quantitative descriptive analysis, *Journal of Food Science*, vol. 57, issue 5, pp 1168–1184.

Sidel, J, Stone, H 1993, The role of sensory evaluation in the food industry, *Food Quality and Preference*, vol. 4, issue 1–2, pp 65–73.

Stampanoni, C 1994, The use of standardized flavour languages and quantitative flavour profiling techniques for flavoured dairy products, *Journal of Sensory Studies*, vol. 9, issue 4, pp 383–400.

Stone, H, Bleibaum, R, Thomas, H 2012, *Sensory Evaluation Practices*, Fourth Edition, Elsevier, Oxford.

Thorngate III, J, Noble, A 1995, Sensory evaluation of bitterness and astringency of 3R(-)-epicatechin and 3S(+)-catechin, *Journal of the Science of Food and Agriculture*, vol. 67, issue 4, pp 531–535.

Tomic, O, Nilsen, A, Martens, M, Naes, T 2007, Visualization of sensory profiling for performance monitoring, *LWT: Food Science and Technology*, vol. 40, issue 2, pp 262–269.

Tomic, O, Luciano, G, Nilsen, A, Hyldig, G, Lorensen, K, Naes, T 2010, Analysing sensory panel performance in a proficiency test using PanelCheck software, *European Food Research and Technology*, vol. 230, issue 3, pp 497–511.

Urdapilleta, U, Ton Nu, C, Saint-Denis, C, Huon de Kermadec, F 2001, *Traité d'évaluation sensorielle: Aspects cognitifs et métrologiques des perceptions*, Dunod, Paris.

Van Trijp, H, Schifferstein, H, 1995 Sensory analysis in marketing practice: comparison and integration, *Journal of Sensory Studies*, vol. 10, issue 2, pp 127–147.

Vannier, A, Brun, O, Feinberg, M 1999, Application of sensory analysis to champagne wine characterisation and discrimination, *Food Quality and Preference*, vol. 10, issue 2, pp 101–107.

Varela, P, Ares, G 2008, Sensory profiling, the blurred line between sensory and consumer science. A review of novel methods for product characterization, *Food Research International*, vol. 48, issue 2, pp 893–908.

Villanueva, N, Petenate, A, Da Silva, M 2000, Performance of three affective methods and diagnosis of the ANOVA model, *Food Quality and Preference*, vol. 11, issue 5, pp 363–370.

Westfall, P, Tobias, R, Wolfinger, R 2011, *Multiple Comparisons and Multiple Tests Using SAS*, Second Edition, SAS Publishing, Cary, NC.

Williams, A, Arnold, G 1985, A comparison of the aromas of six coffees characterised by conventional profiling, free-choice profiling and similarity scaling methods, *Journal of the Science of Food and Agriculture*, vol.36, issue 3, pp 204–214.

Williams, A, Langton, S 1984, The use of free-choice profiling for the evaluation of commercial ports, *Journal of the Science of Food and Agriculture*, vol. 35, issue 5, pp 558–568.

Worch, T, Lê, S, Punter, P 2010, How reliable are the consumers? Comparison of sensory profiles from consumers and experts, *Food Quality and Preference*, vol. 21, issue 3, pp 309–318.

Zamora, M, Guirao, M, 2004 Performance comparison between trained assessors and wince experts using specific sensory attributes, *Journal of Sensory Studies*, vol. 19, issue 6, pp 530–545.

3

The Foundations of Consumer Evaluation

3.1 Qualitative Consumer Studies: When We Are at the Stage of Proof of Concept

3.1.1 When to Take a Qualitative Approach?

The best way to understand consumers and what drive their decisions is to listen to them. Hence, the objective of qualitative research is 'exploration'. Its main function is to gain understanding on underlying opinions, habits, expectations, motivations or barriers. It provides insight on certain questions, areas or problems and allows to develop hypothesis for further improvements. Qualitative consumer studies are always a small-scale approach that aims at uncovering trends, deep dive into a subject and get elements to grasp the bigger picture. Results cannot be generalized to a larger population but lay out the foundations for potential quantitative research (see Section 4.2). They allow defining postulates to be further validated with quantitative approaches (Calder 1977). In some cases, people conduct hybrid 'qual-quant' studies where some respondents are pulled (randomly or according to certain quotas) from a large-scale study to go more in-depth through narratives, picture collections or observations. In those cases, the qualitative approach is used *a posteriori* to better refine findings of the quantitative study. In all cases, qualitative research should never be used for statistical projections or decision making. In very specific cases, when multiple qualitative tests are combined and findings integrated, conclusions can lead to some decisions with lower grade risks, but even in those cases, high caution should be used (Hall & Rist 1999). Table 3.1 gives situations that justify a qualitative research to be undertaken.

Numbers 1 and 2 are primarily going to help eliminate major defaults or set aside non-viable paths. It is important to fine-tune a target, a context or a method but caution needs to be used in order to avoid missing an opportunity under false negatives. For example, a very promising product could be rejected due to the halo effect of a fragrance, inappropriate packaging or bad instructions. In some

Consumer and Sensory Evaluation Techniques: How to Sense Successful Products,
First Edition. Cecilia Y. Saint-Denis.
© 2018 John Wiley & Sons Ltd. Published 2018 by John Wiley & Sons Ltd.

Table 3.1 Reasons to undertake a qualitative research.

Objective pursued	Results obtained
1) Pilot testing of operational aspects before a larger survey or product placement	Check wording of questions
	Check relevance of a whole questionnaire (fluency and length primarily)
	Check wording of instructions (written, visuals or videos)
2) Proof of concept testing	Verify viability of a prototype (product and/or packaging)
	Verify how a concept, a brand or an idea is perceived
3) Exploring an unknown domain	Define hypothesis on habits, attitudes, behaviours, needs met or unmet and expectations
	Understand consumers or differences among specific consumer segments
4) Generating ideas	Brainstorm around a theme for a creative purpose
	Uncover market opportunities
5) Qual-quant diving	Pull respondents from a large quantitative study to further dive into certain aspects

cases, this approach can also open new perspectives not previously foreseen before. For example, a product could be tested on a group of women with curly hair and give poor results, but among the sample, one woman mentions her husband used it for fun and that '*he actually liked it*'. In a qualitative research, any cue is to be taken with interest. In the same way, a default mentioned by one participant needs to be observed with a magnifying glass as it can turn into an exponential major issue on a larger scale. Thus, the difficulty of the qualitative research will rely on the analysis to ensure not to miss any point, not to amplify anything wrongly.

Numbers 3 and 4 are more open approaches whose objective is to gather consumer insight on an unfamiliar area or a poorly known consumer segment (see Section 1.1) or to boost a creative process. Some companies use this approach in the form of innovation workshops with internal or external participants. Sometimes it can be done with professionals in a specific category (e.g. chefs to create new recipes).

Lastly, number 5 is used to deep dive into certain results of a quantitative study. Sometimes, from the start, it is anticipated that more in-depth information might be needed on certain aspects. This may be solely for illustrative purposes. For instance, some pictures before and after using a product can be taken (for a visual effect on a hair product or a make-up product), or some samples can be taken (to assess how much product was used, to see how a certain packaging resisted certain conditions, samples can be recuperated). But sometimes, it is anticipated that certain responses will trigger off the need to dive deeper into the meaning and implications. In the latter, it can be decided to ask, systematically or with defined quotas, some participants to pursue the test with an in-depth one-on-one interview or to join a focus group. It is always better to anticipate and decide on a hybrid qual-quant right from the start as it can be challenging to trace participants from a quantitative study willing and available to participate in the second phase study.

Very often today, marketers tend to despise qualitative approaches and turn preferably towards big numbers. However, it is important to take a second look and value what more human and small-scale approaches can bring as they tend to tap into the hearts and heads of the consumers. Data mining, in which big data is processed to uncover trends, is valuable. But qualitative approaches often allow to search for areas of organic growth in the very multidimensional and complex consumer's world (Belk et al. 2013, p.5). Hence, qualitative methods tend to resurge as a way to brainstorm, dive into intuitions and ultimately develop better concepts.

3.1.2 Define the Test Design: With or Without Product Testing

Defining the test design rigorously requires going through the following steps thoroughly (Figure 3.1):

- Define the needs and therefore the objectives of the research (Mariampolski 2001, p.65). This will allow to develop a **brief** to frame the project and the output that is expected
- Test **plan** for the project based on that brief:
  - Products to be incorporated in the test
  - Geographic location(s)
  - Consumer target(s)
  - Sample size
  - Timing (time of the day, season for fieldwork, length of the field work, time required to prepare field work, time required to complete the analysis and report).
  - Define the budget to run the study.
  - Elaborate a screener for recruitment.
  - Elaborate a discussion guide for all steps of the project.
- Elaborate the template or outline of the final report or **outcome** (which can be a text document, a set of slides or a video in some cases). Final report template and discussion guide are closely linked.

In some cases, as stated in the previous paragraph, the objective is to explore an unknown domain or generate ideas. This may not always require the use of any product during the test but rather to recruit a specific audience: users of a certain product, of a certain brand, people who have certain habits or routines, a specific age group or people living in a defined geographic zone. For example, we may want to dive into social media habits of college students, we may want to explore smartphone apps used by young professional women or consumption of cereal bars by teens. Thus, recruitment will be key to focus on the precise target audience. Also, no product may be required when the objective is to verify viability of a concept or to pilot test a questionnaire prior to a larger survey that does not involve product testing. But here again, it is necessary to recruit the right target that represents what the larger audience will be.

Figure 3.1 Process to design a test.

Conversely, when the viability of a product or prototype is being assessed, one or multiple products will need to be placed in the hands of the consumers at some point during the field. Different scenarios are possible:

- Products are handed out and consumers can use them for a defined period (at home or at a central location), prior to a one-on-one interview or a focus group. After the period of product use the product can be rendered or not; eventually the product can be weighed before and after if it is important to know accurately how much was actually used. This approach (illustrated on Figure 3.2) is relatively common for well-established product categories where just a new variant is tested (a classic shampoo or body moisturizer for instance). It is very important to be aware of certain incompressible step lengths (such as 2 weeks for recruitment usually recommended or 2-weeks minimum to complete a final analysis) to deal with pressure from management who often looks to expedite studies which can result in quality being at risk.

  For long field periods, the mid-point follow-up may be important to assess whether the product is being used correctly, whether there are any issues or whether the consumer has enough product.

- In some designs, one-on-one interviews or focus groups can happen at different time points during the product experience: after the first product experience, a one-on-one interview or focus group can gather first time impressions. Then the product may be used for a period at home, followed by an interview or a focus group after the home usage period. This design allows to capture the first impression as well as the long-term impression.

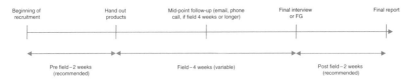

Figure 3.2 Timeline for a classic qualitative test.

It can be interesting for very innovative products. For long fields, the mid-point follow-up can also be a more extensive interview. In all scenarios, if appropriate, an initial interview can be scheduled to talk about general habits and expectations, as well as sentiment on usual product(s). In the latter, if done at home, it is an opportunity to explore the products that are possessed and reasons for choice and usage. If done in a central location, it is recommended to ask the respondent to bring his/her usual products to be more accurate and feed the discussion (based on memory some consumers may just remember the colour or the shape of the packaging but not the actual complete brand name, especially for those that have different variants).

- Some field work can also involve multiple products, in which case for small scale studies it is always recommended to use sequential monadic scenarios (see Section 1.2.1.4).

It is always better to allow the consumers to experience the products at home to be closer to reality. However, for a product to be released, it is important to have no confidentiality limits (some companies may not want a prototype that is very innovative or for which patents are pending to be released for testing). Of course, all safety precautions need to be taken and precise indications need to be given to the consumers on what do to in case of any question or adverse reactions (contact information needs to be reachable at any time).

For very innovative prototypes, one may consider having at least one or two testing time points organized in a way that an observation of the usage can be done, either at a central location or with a home visit.

3.1.3 Define the Market and Consumer Sample: Sample Size, Focus Groups or One-on-One Interviews

Even if it were possible, in a qualitative approach it is not necessary to collect data from everyone in a community; usually only a subset of a population is selected for a study. Hence, sample size (n) in qualitative research is always small, going anywhere from one subject (rare) to 20–30 consumers. Beyond $n = 30$, the terminology often used is 'mini-quantitative' (typically $n = 30$ to $n = 80$). Mini-quantitative tests can be controversial when they are interpreted in a too confirmatory way. Indeed, although the slightly larger sample sizes give a certain sense of confidence, they still need to be considered as exploratory. Beyond $n = 100$, one can assuredly consider being in a more quantitative domain. Different methods can be used to define sample characteristics and size:

- Purposive sampling: participants are selected based on specific criteria.
- Quota sampling: it derives from the previous one but different groups are defined with specific criteria in each one, sub-groups are usually defined to reflect proportions in population.

- Snowball sampling: some people are recruited and refer new participants through word of mouth that match the criteria. It is necessary to be cautious in this case as it is usually recommended not to recruit related participants (friends or family), especially for focus groups, as group dynamic may be strongly impacted and biased.

Sometimes the sample may not be set in stone, but in light of ongoing analysis, lead to new emerging samples to be investigated. Overall, deciding on a sample size and/or quotas is usually not an easy task and needs to be defined on a case by case situation, also often limited by budget and time constraints.

Participants in a qualitative study are typically 'naïve' consumers from the field of interest. In some cases, it can be interesting to conduct qualitative research with 'experts' of the domain that is being studied. For example, some food companies may conduct qualitative research with bakers or chefs to generate ideas or new recipes. Some car manufacturers gather Formula One racers to get input on new car prototypes. Beauty industry often gains opinions from cosmetologists on expectations, needs, gaps or fit of new products.

In qualitative consumer tests, data collection involves individual interviews, group discussions, also called focus groups, or observations (Table 3.2). The term 'in-depth', which is often used, implies that through qualitative research one often seeks information that is more profound than easily accessible via regular interpersonal relationships (MacDaniel & Gates 2005, p.116).

For focus groups, the size of the group that is typically recommended is usually of 6–8 participants (a focus group can be conducted with as few as 3–4 participants up to a maximum of 10). In case the study needs to involve more than eight participants, it is usually recommended to conduct multiple focus groups in parallel and to consolidate results (with the risk that different groups may lead to different outcomes that will need to be analyzed). In many cases, when multiple focus groups are conducted, it provides the opportunity to explore different segments such as different age groups, men and women, and specific occupational links. Although heterogeneous groups yield richer information, most often homogeneity within one group is recommended to facilitate report and to lead to exploitable conclusions (Liamputtong 2011).

Some people use specific terminology that distinguishes focus groups (6–8 participants), mini-groups (4–5 participants), triads (3 respondents), paired or dyads (2 respondents) and one-on-one interviews (1 respondent). The notion of focus group slightly differs from smaller groups in the sense that rather than being a strict group interview, it is more rigorously a group of people engaging in a collective discussion on a topic disclosed by the moderator.

The decision on whether it is better to orient the design towards individual interviews or focus groups needs to be based on whether it is more important to have a range of opinions or to benefit from a group dynamic, 'group-thinking'

Table 3.2 Means to collect data.

Type of collection	Location	Description
Observation	CLT (one-way mirror or video camera recording the happenings) or on-site (home, workplace) discreetly in the background	The researcher watches the subjects and takes notes. Gestures and expressions or emotions are observed. Quantities used can be weighed as well, for example. Depending on the setting, the researcher-observer can often be part of the situation being studied.
One-on-one interview	Face-to-face, phone, webcam and popular videoconferencing apps	Individual interviews are necessary when the objective is to collect personal unbiased opinions on a product or a subject. It allows to go more in-depth into emotional aspects. It can also be preferred with sensitive subjects (personal hygiene routines for instance). Although, for some sensitive topics, people may feel more comfortable sharing experiences among peers (Frith 2000).
Focus group	Traditional FG room, or via webcam w/ synchronized or asynchronous group discussions (online bulletin board where participants connect and respond to questions from the moderator over a few days' time frame. Participants may or may not see what others are responding).	This technique is preferred when a range of opinions is needed and/or when the dynamic of a group discussion on a subject or product will better help define how to move forward or boost a creative process (Sztainer et al. 1999).
Hybrid studies	Individual interviews can be followed by a focus group confrontation of the respondents or after a focus group, all or some of the participants can be interviewed individually	This approach may be burdensome but can be justified if, for example, subjects are to use a product at home. It can be important to have a one-on-one interview with each one on what they thought about the product, followed by a focus group confrontation (the focus group can add some brainstorming opportunities on how to improve the product, what to call it, which could be rather poor individually). As if only the focus group is done, individual opinions of some more reserved participants may go unnoticed.

In some approaches, consumers may be gathered for a focus group on a general topic (usual routines on certain domains, needs, gaps) and then be given a product to be used at home, followed by an individual interview. |

and interactions among participants to boost conversations (Smithson 2000). It is important to keep in mind that in focus groups individuals have less time to speak and some may refrain, hide or be more passive. Focus groups have the advantage of the rapidity with which data can be generated which is often valued.

Qualitative research offers an infinite number of possibilities as any aspect of the design can be customized to meet the needs of the brief. Being an exploratory approach, any design needs to be thoroughly thought in a very open and practical state of mind (Bryman 1988). Furthermore, today, even the frontier between qualitative and quantitative methods may not be as drastic as their definitions may suggest. Many teams have developed in a very pragmatic way, mixed methods to fulfil the needs of their research (Brannen 2005; Williams & Vogt 2011). And if qualitative thinking is always a good starting point multidimensional 'qual-quant' view angles may sometimes be a good strategy.

The explosion of online tools also offers an infinite number of possibilities to outgrow the traditional in-person observation. Cell phone video cameras, regular digital cameras consumers have on their desk or laptops allow nowadays to keep track and even record what is happening in the consumers' intimate world. It is now relatively easy to see how consumers use a product or interact with any element they are shown. This is particularly true with millennials and centennials who are usually inclined to take pictures or videos of themselves at every single moment of their day. Online bulletin boards or communities can be created to collect data remotely much more easily and cost-efficiently. Online tools now offer the immense advantage of getting deeper into consumer's life at a very experiential level at hours' way outside the traditional business hours range and in more private moments. It also allows much more cost-efficient tests (Stewart & Williams 2005). The lower cost to set up studies can however be sometimes counterbalanced by the mass of data (photos, videos) to analyze if they are not sufficiently limited in number. The variety of the material collected (format and content) can also be a difficulty to overcome.

The discussion guide is always designed to stimulate participants to produce as much as possible unguarded, spontaneous comments. The creation of rapport between participant(s) and moderator is key. However, it must be done in a non-bias way as the main critic of this type of approach relies on the concern that they can be very subjective and any given result can be different with a different moderator (in the same way that it can be different with a different respondent or a different setting). Being able to conduct a non-bias qualitative interview or to successfully moderate a focus group requires a solid psychology background to avoid any influential comments and to understand and analyze human actions (participants' observation) and meanings (written or most commonly recorded narratives) (Fischer 2006). Being constantly attuned, alert and attentive to what the interviewee is telling or not is fundamental. Keeping an open mind to different points of views, being able to sense feelings and emotions, being aware of context impact and assuming knowing nothing to begin with to see exclusively from the consumer's point of view are the core skills that are required (Mariampolski 2001). Ultimately the researcher seeks to evoke responses:

- Meaningful and culturally salient to the consumer
- Unanticipated
- Rich and explanatory in nature

Extremely valuable information on how to conduct qualitative interviews are given by Edwards and Holland (2013). The researcher conducting the interview must be flexible and, even though the discussion guide has a certain flow of questions, it is necessary to listen carefully to the participant, bounce back asking why, how, engage considering style and personality and 'probe' accordingly. It is not uncommon to skip parts from the discussion guide taking the opportunity of a specific remark from the participant to then return later to the omitted questions. This flexibility does not mean that there is a lack of structure in the interview. Often, the term 'non-directed' interview is used to signify that despite the structure of the interview guide, the discussion follows the flow of the participants as much as possible, with re-directions if the flow goes off-topic. Indeed, basically the researcher has topics, themes or issues that need to be covered but with a fluid and flexible structure (Hammersley 2013).

The discussion with the consumer usually takes place with the following steps:

- **Introduction to set the stage for the discussion** (once consent forms are signed and the recording device is working – if applicable, discussion can start): who is everyone (it is very common to start by general questions such as passions or hobbies), why is everybody gathered? What is the study about? Why is the interviewee's opinion important? It is necessary to avoid a situation where the participant feels exposed or intimidated, the ideal situation being when he or she feels fairly engaged and valued.
- **Listening, probing, following up, reflecting on what is said, formulating next question, shift to a new terrain when appropriate**: concentration and effort required from the moderator are relatively intense to cautiously make each decision on how to intervene and keep an eye on timing. Topics usually covered may broadly include but are not be limited to:
  - Usual routines and context
  - Focus on test product
- **Wrap up/conclusion and perspectives**: round table, focus on future, review of experience, of most important things discussed. Thank interviewee and keep listening. It is not uncommon that participants open up again with interesting information once discussion is 'officially' over.

Discussion guides are generally used to ensure that certain important topics and issues are covered and are always made of a succession of open-ended questions (as opposed to closed-ended questions primarily in quantitative questionnaires). Typically, individual interviews or focus groups last between 1 and 2 h.

The art of probing, described thoroughly by Bernard (2000), involves multiple techniques such as:

- **Silence**: provides participant with time to reflect, can be difficult for the interviewer but effective when used appropriately.
- **Echo**: consists in repeating last statement from the participant, shows they are understood and encourages to expand.
- *'uh-huh', 'yes', 'I see', 'right ...'*: affirms what the participant says and encourages to continue.
- *'Tell me more', 'what did you mean by..', 'Why do you feel...'*: acknowledges what is said and encourages to go further.
- **Long question**: a very long detailed question can help in the beginning of an interview or when the discussion is blocked. *'Tell me about your morning make-up routine, at what moment do you start, where do you start, what do you use first?'*
- **Leading**: leading questions should be used with caution, usually as direct probes. *'You mentioned doing this is not a good idea, do you really think it is bad?'.* If used appropriately, the answers are not necessarily biased as the interviewee can bounce back differently.
- **Baiting**: this technique can be used, for example, on a topic that people may be uncomfortable to mention, in which case the interviewer can act as if they already know, which can encourage the interviewee to open.

Bernard (2000) also gives very beneficial advices on how to deal with very verbal or very quiet participants. With very vocal interviewees who tend to go off topic, the interviewer should not hesitate to interrupt in a respectful way. As of excessively non-verbal participants, sometimes putting an end is the wisest thing to do without feeling it is a failure.

Appendix 3.1.3a gives an example of one-on-one discussion guide (with product test involved)

Appendix 3.1.3b gives an example on a focus group discussion guide.

Sometimes a <u>stimulus</u> can be added to the protocol as a starting point (word cards, pictures, magazines, videos, games, collage exercises). Some use methods called '<u>projective techniques</u>' where participants are put in a situation to 'project' what they really feel or think. For example, they can be asked to view a short video and fill in the dialogs or comment on what is going on as an open-ended elicitation. The idea is to have them project their opinions in situations where people may not be willing to discuss topics spontaneously.

Literature found on qualitative interviews often mentions other techniques such as laddering, which may in some situations help gather information on consumer deeper perceptions (Miles & Frewer 2001; Roininen et al. 2006). In the laddering technique, questions are organized in a way that the interviewer will make the interviewee name and describe each tangible product feature (e.g. 'Which aspect do you like the best?'). For each one, subsequent questions will form a chain that allows to associate functional tangible benefits (e.g. 'Why? What does it do?'), emotional more subjective benefits (e.g. 'Why is that important to you?') leading to personal values and goals linked to it (e.g. 'Why? Why do you need that?'). This type of progression in the way the questionnaire is conducted ultimately allows to tap into deeper meanings for the consumer. Basically, it consists in probing and digging until personal values are uncovered. The drawback is that interviews can be rather tedious to conduct and long to analyze as the output will consist in visual chains of feature-functional benefit(s) – emotional benefit(s) – value(s) weighed by relative number of respondents who follow that pattern (Veludo de Oliveira et al. 2006).

To deep dive into emotions that consumers may experience while tasting or using a product, another technique is sometimes mentioned called emotional curves, emotional journey or temporal dominance of emotions (TDE) (Jager et al. 2014). Indeed, emotions often evolve during the different stages of the product experience (Schifferstein et al. 2013). To capture the dynamics of those emotions felt along the 'product journey', participants are given a timeline on which they are asked to represent how their emotions vary over time. The information can be captured as a continuum of temporal dynamics or as single ratings at different time points if timeline is broken down into the different steps of the product experience (example for a yoghurt: initial visual aspects of packaging, opening of the cap, visual aspects of the product, aroma before consumption, breaking down the product with a spoon, first spoonful, following ones, last one, after taste).

In the structured timeline, for each time point, consumer is asked to give a rating globally to his emotional state. The scale can go from below zero values (representing negative emotions), to above zero values (representing positive emotions) (Figure 3.3).

After the participant is done tasting or using the product and rating his emotions for the different stages, the interviewer can deep dive into the reasons and annotate his grid with words and comments, more or less rich and elaborated (Figure 3.4).

This method presents the advantage of giving a visual support for the discussion. However, it does not bring a significant advantage compared to a standard in-depth interview were the different steps of the product experience are explored. Nevertheless, it can provide a visual for a presentation of results as well. Indeed, it can be more convincing for an assembly to see visually how the consumer feels over time rather than only going through selected verbatim.

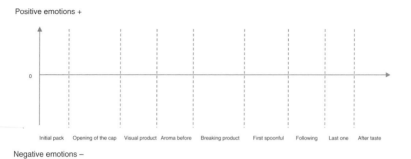

Positive emotions +

0

Initial pack   Opening of the cap   Visual product   Aroma before   Breaking product   First spoonful   Following   Last one   After taste

Negative emotions −

Figure 3.3 Frame for emotional journey log.

However, it is only easy to consolidate if the different participants show a relatively common pattern which is not often the case. Also, for this method to provide significantly richer results, it is necessary to recruit talkative consumers that can articulate beyond purely descriptive vocabulary.

Jager et al (2014) explored food-evoked emotions over time in a more complex way, as they decomposed emotions in 10 different attributes and had the participants assess them over time using a sensory time-intensity type of method (see Section 2.3.5). In parallel, they also assessed Temporal Dominance of Sensations (TDS) for 10 sensory attributes (also called Time Intensity T-I, see Section 2.3.5). In their study focusing on dark chocolates, a complex multivariate analysis was conducted and revealed significant differences based on dominance duration of sensory and emotional attributes. This type of approach may be interesting as a complement to traditional sensory profiling to characterize products or formulas more in-depth incorporating emotional dimensions. However, the data collection requires a complicate logistics (appropriate software to capture data, like time-intensity tools) and thorough multivariate data analysis. Some have tried to have participants draw continuous emotional curves manually on paper grids. This usually does not work very well. Very often participants start drawing a curve, realize they went too far in time and want to erase. Ultimately, the forms obtained are sloppy and hard to consolidate and interpret.

Another recent method is often used to dive into consumers' emotions in the context of using a product; it is the think aloud protocol where consumers are asked to express everything that comes to their mind, either concurrently or retrospectively (Van Den Haak et al. 2010). This approach was originally developed to analyze problem resolution processes and is a very thorough way to deep dive into consumers' minds when they are going through a step-by-step product usage process (Fonteyn et al. 1993). However, some raise the concern that, having the participants think aloud may trigger changes in their cognitive

Positive emotions +

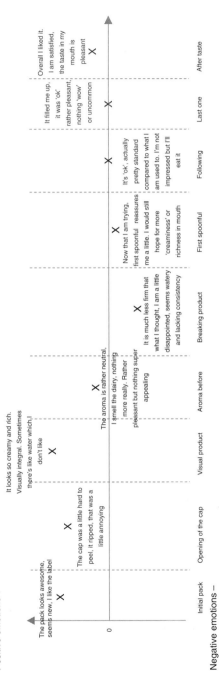

Negative emotions −

Figure 3.4 Example of completed emotional journey log.

processes (Leow & Morgan-Short 2004). Hence, think aloud is a very interesting approach to have consumers speak spontaneously and gather unconstrained thoughts and eventually deeper emotions. However, it may not be combined with a more traditional questionnaire afterwards as the consumers are put into a situation where they may deepen some thoughts much more than in a real-life situation.

Also, often logistics are difficult to put in place and that may result in higher budgets. The easiest way is to ask the consumers to write down their thoughts on a blank page. Or the protocol can encompass a recording or an interviewer taking notes. For some products, to be used outdoors or under the shower for example, it might be very difficult to record concurrently and the thoughts may be gathered retrospectively, which can imply a certain globalization of the overall experience and an impoverishment of the content. Some have experienced a remote recording, via regular phone put on speaker or via video conferencing tools. Overall, this approach can reveal very in-depth aspects for some product categories or stay very descriptive for some others. Indeed, there are product categories that consumers apprehend in a very practical way where the emotional content that is collected can be very poor. When that happens, it is not necessarily a failure of the method but a possible proof that the product category does not reach emotional spheres in the consumer's mind.

Going a little further in assessing emotions, more and more research is conducted nowadays on how to 'objectively' describe and measure emotions experienced by consumers (Laros & Steenkamp 2005; Jessen & Kotz 2011) as well as how emotions relate to acceptability (King & Meiselman 2010; King et al. 2010). This complex field is still being explored; most methods used evolve around approaches where consumers rate selected emotions attributes, as well as facial or vocal expression analysis (visually or digitally) or physiological measurements (e.g. heart rate, blood pressure). However, the precise links between physical or physiological manifestations and specific emotions are still not fully established. Most advocate for combined multi-channel approaches to gain accuracy in predicting models that are established (Desmet 2005).

3.1.4 Define a Timeline

As described generally in Section 1.2.4, framing the study schedule and timing is primordial. Section 3.1.2 goes over the study timeline which defines how long the study will take from start to finish. Furthermore, many products can be tested all year round and/or at any moment during the day. That usually makes planning easier. However, there are many products that consumers use at specific moments throughout their daily routines. When products are tested at home, consumers will usually adapt and use the products when convenient and usual for them. If test happens in a CLT, it is important to schedule the visits accordingly. Often this imposes certain constraints and impact the general timeline. For example, men testing a shaving cream may only be scheduled during a short time frame in the morning. Women testing a night cream, may only be convoked at night. Some products are used daily, some may be much more spaced in time. Women will only colour their hair once a month or less. All these considerations are important as they impact the organization and therefore the cost to be anticipated.

Lastly, some products may be seasonal. Women with curly hair may straighten more often in the winter compared to summer, especially in regions with humid summer weather. Therefore, it may not be a good idea to test a product to straighten hair in the summer as it is not usual practice, results may be biased and recruitment may be more challenging. Indeed, it will be difficult to find women who agree to do it if it is not common for them. In some cases, they may agree for the compensation; however, results may be taken with caution as they will not be doing something they would typically do under those conditions. In the same way, in regions where weather is very cold in the winter and common practice is to wear hats, it may be better to postpone a hair spray or gel test to a later time. In most regions, consumption of ice-cream, frozen yoghurts or smoothies may also be very seasonal as well. Those are example that show how important consumer insight on habits needs to be incorporated in any test design, especially small scale where any bias may be amplified.

3.1.5   Analysis and Deliverables

In qualitative consumer studies, the data analysis consists on the coding of text data (interview transcripts, see example on Table 3.3). During the field work, interviews can be recorded discreetly with an audio recorder. Audio recorder presents the advantage of being able to go back and re-assess emotional wording, emotional timber and tone as needed. However, it may be very time consuming. Some researchers prefer to take notes directly on a copy of the interview guide (paper or computer). Recording can include video but this needs to be done with caution not to be too intrusive (Belk et al. 2013, p.120).

Thus, responses to the open-ended questions constitute a textual content (non-numerical data) that needs to be categorized to become understandable and to be synthesized. Researchers in this field need outstanding analytical skills to identify themes, patterns and ideas. Then, words and expressions are aggregated in clusters based on the analyst's interpretation. Occurrences are counted to define the relative importance of the different trends and implications (Belk et al. 2013, p.139). However, there are no statistics involved. The analyst will also need to focus on grammar, word usage and underlying

Table 3.3 Example of interview transcript (final wrap-up comments on an anti-wrinkle cream).

This consistency is new to me. It is luxuriously creamy. It is much thicker than the normal creams I have most recently and consistently used. I just love this thicker cream, especially at night. I am 53 years old, I am often in the sun and I want to look good for my age. My neck and wrinkles have benefited from this a lot.

I like the results. The cream goes on smooth, really smooth. You really do not need a lot; a little amount goes a long way. I feel it tightens my face rapidly while still feeling hydrated and refreshed. My dark circles got lighter and my neck feels firmer too. I got compliments at work! I had a natural glow too. It is non-greasy, nor glossy. My face felt hydrated, I felt younger!

This felt really great! It is very thick and luxurious. It really hydrates my mature skin. I noticed a softer and smoother appearance.

After trying this cream, I am very sceptical. My face felt dry. It is not rich enough for what I need, it lacks some moisturizing effect. Once I spread it, it disappears almost instantly, almost feeling as I put nothing on.

Table 3.4 Classification of numbers for a qualitative study (example of a $N = 15$–16 study).

Number of occurrences	Terminology
15/16	All
13/14	Almost all
9, 10, 11, 12	Over half
7, 8	half
4, 5, 6	Under half
3	A few
2	A couple
1	Single mention

messages (Spiggle 1994). This analytic method is called the 'coding process' which can go from broad initial coding to more selective narrower categories.

One practical traditional technique still often used, is to transcribe all verbatim into a large excel chart. One column may be used per open-end question. Then highlighters or any colour coding tool may be used to code words or expressions that are to be entered in similar categories. Table 3.4 shows how numbers can be classified in a small sample to weigh the relative importance of the different dimensions pulled out from the data.

Let us say in a study for a face moisturizing lotion, fifteen women are interviewed after using daily the product at home for 2 weeks. Out of the

15 women, 15 mentioned the cream was thick with a rather negative connotation with verbatim such as:

> *The consistency is very thick compared to what I am used to*
> *The cream was thick and harder to spread than usual*
> *The lotion is a bit too thick*
> *The cream is very dense, even a little viscous*
> *The lotion is heavy*
> *When I start spreading it is a bit chunky and thick*
> *The cream is bulky*
> *This product is pretty consistent*
>
> ...

Evidently the report will state: '**All** women in this study stated...'. It is important when analyzing the verbatim to keep track of what point in the interview each statement was made and to encompass that context to categorize and aggregate. Indeed, when one woman states '*the lotion is heavy*', if that statement is given in the context of exploring how the cream looks and feels at the beginning of the product experience (while dispensing and first starting the spread it), it most likely means something similar to what another woman calls '*thick*' at that same time point. However, if the term '*heavy*' is used to describe how the cream feels once applied/partially absorbed or along the day, the significance may be totally different. In the example above, after the primary coding, it may be important to dive into a more specific categorization of the terms that are used to describe the cream. Indeed, it undoubtedly conveys a negative connotation for the fact that it is perceived as thick, but some women use more specific terms (and most likely deeper unpleasantness) that are important to report to give guidance to the product developers on aspects to improve in the formulation ('*chunky*', '*bulky*', etc.).

Let us now imagine that four women mention, in a way or another, that they liked the colour of the lotion, with two women describing it as pearly and saying it looks luxurious:

> *The colour is pleasant*
> *The lotion is white, I like white for face lotions, I am used to it*
> *The colour looks pure, I like the pearly aspect, it seems luxurious*
> *There is some sort of shiny pearly aspect that makes me think it is a*
>   *premium product*

These four statements can be regrouped into a positive colour category. The report will state something like: '**Under half** women commented positively on the colour of the cream'. However, it seems obvious that the first two comments, although positive, show less enthusiasm than the two following ones.

Therefore, a further statement may add: '**a couple** of women describing positively a pearly aspect which infers a high-end product'.

Furthermore, there are also several softwares on the market that allow automated or semi-automated computerized processes, especially since more and more researchers are looking at online conversations and online reviews that tend to generate very large amounts of data. Therefore, data mining techniques, originally more used for quantitative data tend to be exported to qualitative data (Belk et al. 2013, p.93). The exponential progress in computing power probably means than in a few short years, a significant part of the transcribing process from a video feed will be fully automated, bringing costs down.

As stated in Section 3.1.3, usually interview guides are structured in a way that participants are asked broadly in the beginning what they think of a product, then, there is a focus on every step of the product experience and then there is a final wrap-up question that typically asks to summarize what are the most outstanding points to keep in mind. It is important to weigh the importance of the positive and negative aspects taking that into account. Indeed, something that is spontaneously mentioned in the very beginning, that the consumer states and describes again during the heart of the interview and brings back again at the end, is definitely something that needs to stand out in the analysis; *a fortiori* even more if multiple participants comment on that. Another aspect that is barely mentioned just when probed during the heart of the interview may not be as salient.

A qualitative report will always include the elements listed in Table 3.5.

3.1.6 Budget Considerations

Cost associated with a qualitative research include:

- Product samples and labels
- Screener development
- Recruitment of participants
- Product placement (if applicable)
- Participant incentives
- Discussion guide development
- Facility rental (venue to be rented for observations, interviews or focus groups or if in-home may require mileage and extra time from the interviewer/moderator)
- Depending on the time of the day, often respondents are provided food or snacks
- Interviewer(s)/moderator(s)
- Full analysis and report
- Video recording when there are observations

Table 3.5 Qualitative consumer test report content.

Sections	Description
Objectives and project brief	Context, why is this research conducted?
Tested product(s)	Detailed product description that are presented to the consumers, with a picture (if applicable)
Sample/audience description	Size, location, socio-demographic characteristics (age, gender, urban or not, professional activity, etc.), their type of product usage, for certain product categories, physical characteristics, like hair type or skin may be necessary
Timing	Dates for field preparation. Field description. Postfield timeline
Methodology	Focus groups or individual interviews (length of interviews or focus groups). Product use with length and context in which product was used (location, instructions, protocol that needed to be followed)
General findings	Cultural context, key aspects learned about the audience and the category. Level of awareness, motivators and barriers, needs and gaps, expectations. Historical background if applicable, evolution of the category, past, present and future behaviours
	Each point illustrated by visuals, verbatim, videos
If product test involved	Detailed likes-positives/dislikes-negatives and areas of improvement for each step of the product experience. Eventually new domains to be explored (potential new target, new protocol, new domain or region, for example, that was not explored but emerged via some comments)
	Each point illustrated by visuals, verbatim, videos
If usages being explored	Focus on consumer language, on how and why things are done, innovation ideas, positive aspects of what is on the market, drawbacks of used products, unmet needs
	Each point illustrated by visuals, verbatim, videos
If a concept is being explored	Definition: how do consumers perceive, receive or define the concept, associations, is there different meanings and expectations depending on the context, on the audience, on the type of product, on the moment during the product experience
	Each point illustrated by visuals, verbatim, videos
Actionable recommendations	Summary of key findings on general context, on tested product(s) with strengths and weaknesses
	Recommendations for next steps, for market opportunities. Ultimately, articulate what a product or an idea must be like to best fit the consumer's needs and preferences
Edited videos of the usage and sessions	In some studies, the final output is solely a video with comments that gather the entirety of the findings in a very visual way

Qualitative research can be conducted internally, a practice that exists at some sizeable consumer goods companies. However, many opt for outsourcing due to resources available or to guaranty impartiality. It is rare to pay under $10,000 for a basic qualitative project. For limited budgets, it is a good idea to explore Universities and Business Schools as some have marketing research programs which can offer the service. Otherwise, to find deemed market-research companies, many directories are available such as Blue Book published by The Marketing Research Association (https://bluebook.insightsassociation. org), the Green Book published by the New York Marketing Association (https://www.greenbook.org/aboutus) or the more specialized Qualitative Research Consultants Association (http://www.qrca.org). Criteria to find the best firm to conduct outsourced qualitative research include:

- Geography: it is often better to choose a smaller but locally implanted firm for their knowledge of local habits and as they usually offer more competitive prices.
- Knowledge of the industry: some firms are more specialized or have more experience in certain product categories.

When studies are conducted in different regions or even different countries, the question of whether the same company is to handle the entire study is often asked. If the same company conducts the study in different regions, the advantage will be that the format of the report and deliverables will be the same, which is often desired for communication purposes. However, the drawback is usually the budget, which gets higher (firms that do not have local networks get higher prices to rent facilities and recruit consumers). Also, when the same firm oversees a study in very different geographical regions, quality may be impacted as local knowledge will not be as refined. On the opposite, if different firms are used for different regions, the synthesis work afterwards will require more effort as local formats may differ largely. But the advantage of having locally implanted and knowledgeable researchers may be worth it.

In all cases, it is always important to get several proposals and price offers before moving forward. For further details on when and how to outsource, see Chapter 6.

3.2 Quantitative Consumer Studies: As We Get Close to Product Launch

3.2.1 When to Move Forward with a Quantitative Approach

Quantitative research's objective is confirmatory as it seeks to describe and confirm phenomenon across large number of participants with statistical techniques to recognize overall patterns. It is important to carry out quantitative tests before launching any new product, idea or service as it gives factual figures and data. Quantitative research is based on:

- Large sample sizes ($n \geq 100$ up to 1000)
- Numbers and percentages collected across the sample (percentages are only expressed when sample size of main or sub-groups is ≥ 100)
- Mostly close-ended questions:
  - Verbal scales such as *very dissatisfied/dissatisfied/somewhat dissatisfied/ somewhat satisfied/satisfied/very satisfied* (progressive) or *not enough/just right/too much* (called just about right 'JAR' scales)
  - Ratings on numerical intensity scales
  - Multiple choice lists (with one choice possible or check all that apply 'CATA' option)
  - Yes/No
- Advanced and multivariate statistical techniques

The objectives are usually to test a message, a brand, a concept or a product in terms of satisfaction, comparison to existing market, assessment of a market size or need state. Thus, quantitative studies are put in place to answer the needs of:

- **Product development**: formulators/engineers want to understand how a new product or idea is going to perform on the current market, what fits consumer needs and what needs to be improved.
- **Business**: marketers want to understand consumers' satisfaction, identify opportunities, gaps and needs that can be fulfilled, allegiance to a brand. Large-scale studies may also allow to break down a wide market space into more specific groups with different needs and gaps to fix. This approach is called segmentation.

Hence, quantitative studies generally provide more statistical elements to eventually generalize results to a larger population and support decisions. However, they need to be planned wisely as they are usually more expensive and time consuming than qualitative research. Quantitative tests are recommended for validation as a final course of action when all green lights of preceding steps have been turned on. (see Section 4.3).

3.2.2 Define the Test Design: One or Multiple Products

The first step to move forward with a quantitative research project is to precisely define the specific need and frame the expected outcome (are we trying to position a new product in the existing market? Are we trying to create a new category? Are we trying to understand how a brand is perceived?). Just as stated in Section 3.1.2, it is necessary to clearly phrase the needs and objectives which then defines the **brief** and therefore the format and content of the report

Consumer	Product
1	A
2	A
3	A
4	A
5	A
6	A
7	A
...	...

Consumer	Product
I	B
II	B
III	B
IV	B
V	B
VI	B
VII	B
...	...

Figure 3.5 Monadic design: absolute measurements.

or **outcome**. Then, the project test **plan** naturally arises encompassing geographic location(s), consumer target(s), sample size, timing, budget, screener and questionnaire. All these elements need to be intrinsically linked to the expected outcome.

Since quantitative research usually comes after multiple steps and sometimes back and forth of qualitative tests, usually the brief is already clear and defined. However, it may have evolved or have been refined throughout the process.

When products are being tested to be compared to one another in terms of performance, a question that is often discussed is whether to test using a *monadic model* or a *sequential monadic model*. Both methods have advantages and drawbacks (see Section 1.2.1.4).

In the *monadic design*, consumers of one cell evaluate one product only; they do not compare the test product to any other product (Figure 3.5). If multiple products are to be compared, they are tested by multiple monadic simultaneous cells and the statistical analysis done after the tests are completed will look at significant differences between the cells for all measures. Sometimes cells can be separated in time; it is very important to ensure those cells remain comparable (sample characteristics, weather). Indeed, this design presents the advantage that new products (new cells) can be added and compared a posteriori.

In the *sequential monadic design*, consumers evaluate several products in a fully randomized and balanced design (Figure 3.6). It is very commonly used

Consumer	Position 1	Position 2	Position 3
1	A	B	C
2	B	C	A
3	C	A	B
4	A	C	B
5	C	B	A
6	B	A	C
...	...	...	...

Figure 3.6 Sequential monadic design: direct comparison.

to compare two or even three products. Usually, discrimination between products that is obtained tends to be more significant than when comparing monadic cells, but product effects may not be the same as in real life where products are not compared in such a way. Beyond three products, it is important to ensure that the length of the study is not biasing the results. It may be ok for a woman to test seven shampoos, one shampoo per day for fragrance assessment purposes (meaning, if she is a daily shampooer, she is going to use and assess one shampoo every day for 7 days in total). However, it may be more questionable to have a woman test seven nail polishes, as this may require a test extended to a 7-weeks period. This can be lengthy and result in lack of interest towards the subject matter in the end. It is always important to use common sense when defining the design and length of the test.

An additional approach, sometimes called '*proto-monadic*' design consists in conducting a sequential monadic test where each consumer tests two products, one at a time in an alternate randomized order. After the second product is tested, a preference question is assessed. Once the test is completed, a monadic analysis is conducted only on the data of products tried first. This is often considered as presenting the advantages of both approaches as it gives absolute values and a direct comparison and a preference evaluation. However, it is important to have a large enough sample so that the monadic analysis is robust. Presenting products simultaneously (e.g. *paired-comparison*) is usually not recommended as it is heavily influenced by the interaction between products that happens. It may allow to detect if one product is better than the other. However, absolute values are not accurately obtainable that way.

There is one specific case where sequential monadic designs may be used with a large number of samples: the **screening of fragrances or aromas** to select optimal candidates for a final product. Indeed, choosing the best fragrance for a personal care product or the best aroma for a food product can be a daunting task given the number of scents that can be available. Furthermore, assessing consumers' preferences may be a difficult task as these may be inconstant and impacted by trends, seasons, cultural aspects, among other factors. Typically, screening of scents is done in two steps:

- **CLT sniff test** (Figure 3.7): Several candidates will be presented to the consumers in a sequential monadic order. The presentation is obviously blind and among the candidates, usually some benchmarks are introduced (those can be scents existing in the market that typically perform well regarding fragrance or it can be top competitors). For each, the consumer will smell (take off the cap of the container briefly and put it back), fill a questionnaire on fragrance attributes (intensity), overall linking, as well as how the scent matches the category which is usually disclosed (Ex: is this scent suitable for a conditioner?). Appendix 3.2.2a gives an example of questionnaire that can be used). Based on these results, two or three best candidates may be selected for the following step, as well as the top performing benchmark.

  Sniff tests need to be conducted in well-ventilated facilities. Protocols usually state that between each sample, participants will breathe fresh air or smell inside their arm to reset their senses. Some provide a jar of coffee beans, which can also be a way to reset senses.

  For food products, a simple sniff may not suffice, in some protocols, participants are requested to taste a small spoonful. In some protocols (e.g. for coffee or cocoa beans, as well as alcoholic products), the participant is asked to spit the product once he/she gets a good sense of the aroma in their mouth. This allows to reduce the fatigue that can be generated. However, with food in-mouth tasting, the number of samples that can be assessed in one session is necessarily lower. It is important to use common sense to determine how many candidates can be assessed in one session (some pre-test sessions can

Figure 3.7 Sniff screening test sequence.

be run to determine it). For purely sniff tests, it is usually recommended not to exceed six to eight samples. For in-mouth tasting, that number needs to be reduced to five or six.

- **In-home use test**: For personal care products, fragrance must coexist with body scents as well as with other products that are part of the routine. For food products, the aroma may be perceived differently when tested alone compared to within a usual meal and in a real-life situation. Therefore, a CLT sniff screening is not sufficient to predict winning scent candidates. After the sniff screening, it is recommended to test the two or three best candidates as well as the benchmark via large monadic cells at home. Length of the home test needs to be determined so that it allows enough time for the product to be fully used in real life generally multiple times (many scents can be pleasant at first and wear out after multiple uses, for example, in cleaning supplies).

Some product tests intended for **claim substantiation** need to be designed in a very standardized way. A claim is a statement about a product that highlights its advantages, sensory or consumer perceptual attributes or strengths, or product changes or differences compared to other products. Claims that need to be properly substantiated with a valid scientific approach are statements made on advertising media or packaging such as:

- Nutritional content, physiological benefits or the benefits of 'active' ingredients
- Performance characteristics
- Advantage(s) versus a reference – generally competition (meaning taste superiority for foods, ease of application, longer lasting or felt benefits for cosmetics)

It is always important, before deciding to conduct a consumer test for claim purposes, to ensure that there is no scientific measurements or existing data that already provides support for the claim, or that on the contrary no existing opposing/conflicting data exists. Indeed, the rigor that is needed implies that a consumer test can be more expensive and time consuming than an instrumental measure and many claims can be supported by the scientific data that the company may already have. The review needs to be done via a discussion between marketers (who define the claims that they wish to do), scientists (who know how the product should perform) and lawyers (who know regulations and requirements to be able to use a claim without being challenged – or more realistically with enough ammunition to sustain a good enough defence) to confirm the need for a consumer test for each individual claim. Proving physiological benefits (and not just a feeling or perception) usually falls into specific scientific designs and are typically referred as 'clinical study' which obey a specific set of rules – and sometimes regulation – depending on the targeted end-point and outcomes. For any other claim that requires a 'human' detector on more innocuous outcomes such as taste test ('this new cereal taste better than...') or expression of a perception ('my skin felt softer after using this cream for 2 weeks...'), pure instrumental measures are not enough, and so a well-designed consumer test is the only way to go.

Typically, claim tests need to be conducted on large size samples (to have enough statistical power) and outsourced (to ensure impartiality). When the intent is to support absolute, non-comparative performance, the blinded product is to be tested via a large monadic cell and *agreement scales* are the norm to support descriptive or hedonic attributes (see below scales details). When the objective is to prove superiority, unsurpassed or equivalence, a blind sequential monadic design is to be used to support discrimination. For claim tests, questionnaires are usually short and focused on the attributes that need to be substantiated. Reports are also focused on those objectives stating whether the claim is substantiated or not based on statistical analysis. Very detailed information and recommendations on how to select and recruit consumer samples, select and prepare products, questionnaires, execute and analyze the data based on the objective can be found in ASTM (2016). Some claims, in the cosmetics industry in particular, may require assessments by trained professionals (cosmetologists, dermatologists). More details can be found in the manuals edited by Aust (1998) or Ennis and Ennis (2016).

In large-scale studies, the way the **questionnaire** is designed will of course affect the answers. Survey components include:

- The language
- The order of the questions
- The scales that are used for close-ended questions.

There are many types of **scale** options depending on the objectives and the attribute that is measured. It is also very important when comparing multiple studies (e.g. on different cells, conducted in different regions or on different demographic segments) to keep questionnaires and scales identical). A few examples are shown below.

Intensity rating scales:

Linear non-structured: *How much did you <u>like</u> the product (left = not at all to right = very much)*

———————————————————————————→

Linear semi-structured: *Please rate how <u>sweet</u> this product is (0 = not at all to 10 = very)*

———————————————————————————→
0   10

Linear semi-structured: *How <u>easy</u> was it to apply the product?*

———————————————————————————→
Not easy at all   Very easy

These, often called 'semantic differential scales', are very similar to the previous ones but use dichotomous words at either ends of the spectrum (weak/strong, useless/useful). This allows to measure more specific attitudinal responses which remain usually negative/positive anchors for specific attributes. For some dimensions, it may be challenging to find the right dichotomous words. Common sense and cultural/language specifics should always prevail to determine those.

Linear structured numerically: *How <u>easy</u> was the process? (0 being not easy at all and 10 being very easy)*

Most common rating scales in this category are 0–10 or 1–10, 1–7 or 1–5 (sometimes called 'Likert' scales). Some have the particularity to offer the option of a middle point, some do not. Section 1.2.3.2 of this manual details how important it is to take cultural aspects into consideration before choosing a scale. Indeed, some have greater propensity to use extremes, in which case it is recommended to use a larger scale. Some have a greater 'central tendency' than others in which case it might be better to display scales without middle point. Visual presentation (horizontal, vertical, line with indents or boxes) may vary as well, depending on the software used to create and edit the questionnaire. Of course, usual practice is to present the larger number as the 'positive' end, and the right end (if horizontal) or the top (if vertical or presented as a drop-down menu) as the positive anchor. It is commonly admitted as well, that variance in results is higher with larger scales; so many prefer Likert scales to limit variance in results and enhance discrimination (McKelvie 1978). It is usually recommended to choose smaller ranges to measure quantifiable attributes and slightly larger ranges for more subjective or 'blurry' measures such as attitudes, feelings or emotions with the risk of increasing the 'noise' in the data.

Often scales may not be presented as a line but just structured with option boxes. Those are usually referred to as non-linear, structured or categorized scales.

Structured by categories or semantically (<u>attribute</u>): *Please rate the level of <u>sweetness</u> of this yoghurt*

Extremely sweet	☐
Very sweet	☐
Sweet	☐
Just detectable sweet	☐
Not sweet	☐

Structured by categories or semantically (<u>liking</u>): *How much did you <u>like</u> this chocolate bar overall?*

I like it very much ☐
I liked it ☐
I neither liked it nor disliked it ☐
I disliked it ☐
I disliked it a lot ☐

Structured by categories (<u>satisfaction</u>): *After using this product, how satisfied are you with the result?*

Very satisfied ☐
Satisfied ☐
Neither satisfied nor dissatisfied ☐
Dissatisfied ☐
Very dissatisfied ☐

It is commonly admitted that consumers usually prefer continuous linear scales as they allow more spontaneity. However, they induce more fatigue in long surveys as they require more focus. Furthermore, non-linear structured scales usually allow quicker questionnaire designs and may be faster to analyze and report. In some cases, finding the appropriate vocabulary for the categories showing the accurate progression may not be obvious. For example, for 'fragrance intensity', should we say: *not perceivable, very weak, weak, moderate, strong, very strong* or *none, slight, moderate, strong, very strong*? In some cases, it may be necessary to conduct a small local survey (can be done internally with employees) to understand implications of different denominations and define the most appropriate ones.

It is important to point out that with human ratings on numerical scales, the constancy of intervals is not necessarily accurate. This means that when asking on a scale of 1–5 how satisfied a consumer is, a 4 does not necessarily mean twice as satisfied compared to a consumer who rates 2. And the difference between 2 and 3 may not be the same as between 4 and 5. However, it is commonly admitted relying on that approximation and use standard statistical tests (means and standard deviations).

Overall, there is no advantage pointed out in the literature between linear and categorized scales on reliability or quality of results. Choices should be made considering practicality and software capabilities. In case of doubt, small-scale test surveys can help define the best path.

There are other scales that can be chosen to fulfil specific objectives.

Agreement scales:

Structured by categories: *With this cream, my skin stays hydrated <u>all day</u>*

Strongly agree ☐
Agree ☐
Neutral ☐

Disagree	☐
Strongly disagree	☐

These are typically used for claims substantiation purposes.

<u>Re-use (or purchase) intent scales:</u>

Structured by categories: *After using this product, would you re-use it?*

Definitely	☐
Probably	☐
I don't know	☐
Probably not	☐
Definitely not	☐

It can happen very often that a product tests well on most attributes, with liking scores and others rated positively. However, these questions do not necessarily show whether the product presents advantages versus what the consumer already has. Therefore, it is usually important to understand if, beyond the positive attributes, the product has a chance to be incorporated in the consumer's routine in addition to or in place of existing products.

<u>Preference scales:</u>

Structured by categories: *After using product A and B, did you prefer?*

A	☐
B	☐
(No preference)	☐
(I don't know)	☐

Preference measure can only be used in the case of comparative sequential monadic tests. There are a lot of discussions among professionals on whether it is better to allow the option of 'no preference' or 'don't know'. Forcing the consumer to disclose a preference may be a little arbitrary and bias the results. Whereas, allowing not to have a preference may be a safe zone and lead to more neutral results. In the latter, common practice is to either split the no-preference half and half between the two products or just not count those. Advantages of one or the other will be detailed in Section 3.2.6.

<u>JAR Scales:</u>

Structured by categories: *Did you think the strawberry <u>flavour intensity</u> in this custard was*:

Much too weak	☐
Too weak	☐
(Somewhat too weak)	☐
Just right	☐
(Somewhat too strong)	☐

Too strong ☐
Much too strong ☐

This measure is typically used when some characteristics may need to be adjusted in terms of intensity. Section 3.2.6 will detail how to represent and analyze JAR results.

Dichotomous Scales:

Yes/No
True/False
Agree/Disagree

These scales offer two diametrically opposed choices with no chance for nuance. The respondent is not offered the option to be neutral. In some cases, there can be a value in the lack of neutral alternative. Indeed, in long surveys, respondents tend to gradually migrate towards the centre of the scales. However, it can create false positives or negatives and most of the time does not provide specific enough insight on the attribute that is being measured.

Ranking Scales:

When multiple products are presented simultaneously, a ranking question can be asked.
    Please rank these five products from your least preferred (left) to your most preferred (right)?

Least Preferred   Most Preferred

☐   ☐   ☐   ☐   ☐

Demographics:

For the collection of sensitive information such as demographics (age, income, etc.) it is common practice to use nominal clusters:

Under 18 ☐
18–24 years ☐
25–34 years ☐
35–44 years ☐
45–54 years ☐
55–64 years ☐
65 or older ☐

Lastly, in large scale consumer studies, it is always possible to include some open-ended questions. It is usually recommended to include them to allow consumers to give further details on questions such as overall liking, overall satisfaction or overall preference. However, time required to analyze the verbal content is non-negligible as it cannot be automated in the same way as close-ended question responses. This results in additional cost for each open-ended question. Therefore, the number of open-ended questions is often limited to 2 or 3.

Appendix 3.2.2b and 3.2.2c lists examples of questionnaires for large-scale monadic tests as well as for sequential monadic tests. Brace (2008) provides a very comprehensive manual on questionnaire design. Usually the order in which the questions are displayed follows the logical path of the chronology of the product use. It is common practice to ask general questions such as overall liking and satisfaction in the very beginning (with eventual open-ended questions to allow the consumer to further spontaneously explain). Then the detailed survey focusing on every single stage of the product experience may be unfolded. This allows a more spontaneous and a close to real-life assessment of overall measures as in the end, the questionnaire itself may have opened the consumer's eyes into aspects that they may not notice spontaneously. Overall, it is important to keep the questionnaire as short as possible only focusing on the questions that fulfil the brief and end-goal. It is also recommended to keep the format of questions and scales consistent to avoid confusions.

Today, there are several options for **data collection** in quantitative market tests:

- **Mail**: it has now become very rare as it implies handling of data entry.
- **Face-to-face**: it is often done at a CLT, in a mall, or the interviewer can make an appointment and go to the respondent's office. It is very rarely done at home for cost reasons. Usually cost of face-to-face is high but response rate, which is a critical aspect, may be higher too. Skilled interviewers may get more information than in self-administrated questionnaires. This method may be preferred if questionnaires contain multiple open-ended questions. It may be necessary if some respondents are going to be selected for more in-depth follow-up interviews.
- **Telephone**: it can be a good compromise to reduce costs compared to face-to-face, ensure a good response rate and quality of the responses, especially with open-ended questions.
  Nowadays, face-to-face and telephone interviewing is most often computer assisted (computer-assisted personal interviewing [CAPI]) via the use of a portable computer where data is entered.
- **Online/web**: it is the most commonly used nowadays. It allows more easily random sampling, visual displays (depending on software limitations), and allows to easily unfold questionnaires with tasks and conditions. It is usually fast to implement and analyze. The only drawback may be that responses to open-ended questions can be relatively poor.

Some methods to engage consumers may be 'hybrid'. For example, an invitation to an online survey can be done by email (with eventually a website to go to or an URL to the questionnaire directly embedded) or phone call. It is also possible to follow up certain online surveys with a phone call if more in-depth information is needed.

3.2.3 Define the Market

Before starting a study, it is important to define where and with whom the study is going to be conducted based on the brief. This leads to define precisely the target audience and therefore to create the recruitment screener. Incidence of specific consumers that are targeted will have a direct impact on recruitment difficulty and cost anticipation (see Section 1.2.3.1). The market is usually defined based on three components.

1) **Geography**: what country(ies), what region(s), what specific zone(s) (cities) are we going to focus on? The precise geographic areas where the study is going to be carried need to consider population density and characteristics, as well as climate implications and ease of reach.
2) **Demographics**: what characteristics need to be included in terms of gender, age, ethnic background, income, occupation, education and household size?
3) **Behaviours**: for some studies, the recruitment may need to be limited to consumers who have certain characteristics (skin, hair for instance), activities, habits or routines, who use certain types of products or certain brands. Often consumers may be classified into 'heavy', 'medium', 'occasional/light' or 'non' users of certain products or brands based on usage frequency as well as loyalty or exclusivity. Some studies focus on consumers who once used certain products but stopped using them for multiple reasons; usually the term 'abandoners' is used.

A fourth component may be considered, called psychographics. This entails consumer's values, opinions, interests and lifestyle in general.

Some studies may be conducted on different segments or clusters of a wider market. Those segments are then precisely defined based on these four dimensions. They may be compared to each other and to the entire population in the final analysis.

3.2.4 Define the Sample: Sample Size and Confidence Level

Defining how many respondents are needed to ensure a good predictability is a difficult question (Kish 1963; Lenth 2012). Of course, the larger the sample the better the accuracy may get. However, the price of a study is primarily linked to the size of the sample, especially if recruitment is difficult.

Typically, experimental designs are balanced and require having enough respondents to ultimately allow all product combinations to be seen an equivalent number of times. Also, the sample selection needs to be done in such a way that the global population is well represented and inferences can be made from the results obtained.

The optimal sample size is linked to how accurate the data is expected to be. Hence, four factors need to be considered as they affect the accuracy of the data:

- **Estimate of the population size**: this is usually a rough approximation (e.g. size of women population within a certain age bracket in a specific region. Based on penetration figures for certain products or brands, rough approximation of number of users can be made).
- **Confidence interval (also called level of precision or sampling error)**: compared to the population values for the measured attributes, what will be the acceptable margin of error? In consumer studies, the acceptable confidence interval is commonly 5%. Thus, if for example we find that 60% of the respondents are satisfied with a product, with a precision rate of ±5%, between 55% and 65% of the respondents are satisfied.
- **Confidence level (also called risk level)**: how confident do we want to be that results fall within the confidence interval. In consumer studies, common confidence intervals are 90%, 95% or 99% for more conservative evaluations.
- **Standard deviation (also called degree of variability)**: this determines the expected variance in the responses. For consumer studies, the number commonly used is 0.5 to be conservative.

Tables can be found in the literature that determine the optimal number of respondents based on these parameters (Israel 1992). Most experimental design softwares also provide it. It is also possible to find multiple sample size calculators online.

For example, for a size of global population of 100,000, based on a confidence level of 95%, a standard deviation of 0.5 and a confidence interval of ±5%, ideally the survey should be conducted on 398 respondents. Often sample sizes for quantitative studies are smaller than that ideal (200–300) due to logistics and budget reasons.

It is also important to consider response rates. Indeed, in most test fields, the number of fully completed surveys compared to those that are sent out is far from being 100%. Therefore, based on the recruitment method and the anticipated response rate, it is always necessary to send out more surveys than the number of expected completed ones.

3.2.5 Define a Timeline

As described generally in Section 1.2.4 and more specifically for qualitative studies in Section 3.1.4, it is very important to determine when the study is going to be conducted. For large-scale studies where products are used at home, generally consumers are going to pick the time of the day that is most usual and convenient for them to use the product and then complete the survey.

What is important is to ensure that the season chosen for the field is relevant regarding the way the product is going to be used. Also, if multiple monadic cells are run separately, it is crucial to ensure that season and weather are the same and/or have no anticipated impact in the way the product may be used or perceived. Even for purely informational surveys, it is vital to consider whether the moment in which the field is conducted is appropriate. For example, it may be preferable to conduct a survey on sunblocks usage during the summer rather than in the middle of the winter. Some studies can be moved geographically as well to ensure a more appropriate climate.

As for timing to complete a large-scale study from start to finish, timeline described in Section 3.1.2 for qualitative projects might be transposed into large-scale studies. It is usually safe to anticipate at least 2 months:

- At least 2 weeks for <u>recruitment</u> of respondents and <u>product placement</u>. More challenging recruitments may require more time. Based on screener characteristics and available data on incidence in the population, it is important to always be realistic to set accurate and sustainable deadlines. Recruitment may indeed be one of the key hurdles that can result in study delays. If some of the characteristics are expected to be challenging, it is important to list those and decide from the beginning which may be the constraints that could be lifted in case respondents are not found at a reasonable pace.
- Two or more weeks of <u>field</u> depending on the product type and its frequency of use. For long field, mid-term follow-up contacts may help maintain a reasonable response rate. In case too many respondents may be dropping down the study, subsequent rounds of enrolment may be necessary. A certain margin may be anticipated in the budget estimate to cover such events.
- Two weeks to obtain a <u>top-line</u> of results (typically frequency tables and main indicators). In most cases, clients are very impatient. It is very common to incorporate a 'top-line' step in the timeline of large-scale studies. It is important to be very specific in what that top-line entails.
- Two weeks to obtain a <u>full analysis and report</u>.

Conducting a large-scale quantitative study can be a titanic task sometimes. To facilitate logistics and time management, Table 3.6 provides a helpful chronological check list of the different milestones that need to be hit for a successful unroll.

3.2.6   Analysis and Deliverables

The most common way to represent attributes rated on intensity scales is with graphs showing average values (standard deviations may be included as well). Significant differences are important criteria in paired comparisons. Figure 3.8 shows an example where the average rating for product 1's overall liking,

Table 3.6 Successful study checklist.

	Task	Description
1	**Test request** *1+ Day*	It is recommended to have the 'client' (internal or external) formalize a detailed request that specifies: • Needs and objectives: Why is the test considered, what is the output expected? What are the main questions that need to be addressed? What is the global market context in which this project is positioned? What are the consumer insight triggers? • Product(s) to be tested or domain to be investigated: detailed description, instructions (if applicable) • Time constraints and deadlines
2	**Meeting involving different stakeholders** *1+ Day*	It is usually recommended to gather all key project stakeholders. These include (but are not limited to) product developers, marketers, consumer insight department, legal department and evaluation researchers. This encounter usually allows to precisely define the brief, set up expectations and agree on strategy, methods and timelines before moving forward
3	**Budget agreement** *1+ Day*	Stakeholders need to mutually agree on the budget and validate it
4	**Product preparation and collection**	Products should be counted and checked: packaging, proper opening and dispensing, no leaks, labels (which should include proper name and instructions as well as contact information for consumers in case of any question or reaction). Food products may need to be labelled with proper ingredient list and allergens according to local regulation
	Time dependent on product availability	Labels should usually be very precise in terms of manufacturing location and date as well as batch when applicable for traceability
5	**Safety guarantees on the products**	Depending on the domain, products should come with the documentation that guarantees they are proper for usage or consumption under the test conditions and that shelf life is compatible with the test length
6	**Screener development** *1+ Day*	Stakeholders need to mutually agree on the consumer detailed target and validate it
7	**Questionnaire development** *1+ Day*	Stakeholders need to mutually agree on the questionnaire and validate it
8	**Respondents recruitment** *2+ weeks*	This task is one of the key ones which can flow very easily or may require opening of certain criteria. It is usually recommended to over-recruit to compensate for dropouts

(Continued)

Table 3.6 (Continued)

Task	Description
9 **Product hand-out** *Staggered based on sample size*	Depending on how spread the geographic delimitation is, products can be mailed or handed out in specific locations (mall, CLT, etc.)
10 **Field** *2+ weeks*	While products are being used, depending on field length, it may be necessary to have some ongoing contacts with the participants to answer eventual questions, ensure their compliance with the instructions and that they have enough product
11 **Data collection** *Staggered based on sample size*	Completion rate needs to be followed up closely, call backs or reminder emails can be considered to boost participation in this key step
12 **Top-line** *2 weeks*	Usually data is numerically captured and some Key Performance Indicators (KPIs) can be rolled out very quickly
13 **Full analysis and report**	Final output with thorough analysis and recommendations is issued as a final step. The format is typically a PowerPoint document
2 weeks	An in-person presentation if often recommended to avoid any misinterpretations, to address questions, to discuss recommendations and next steps with all stakeholders involved

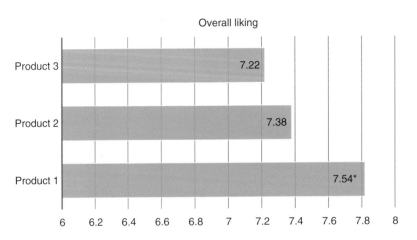

Figure 3.8 Overall liking intensity scale represented with a bar chart. *, statistically significant difference.

on a 10-point scale, is 7.54 and is significantly higher than for products 2 and 3 at a 95% confidence level. Product 2 has an average value slightly higher than product 3. Both have average lower values than product 1. But these two products are not significantly different from each other from a statistical point of view.

Typically, a t-test, also called Student-test (Student 1908), is run to determine significant differences between two datasets (or 'populations') under the hypothesis that both follow normal distributions (Geary 1947). The t-test consists in testing the 'null hypothesis'[1] which, by convention, is that both population means are equal. The two separate datasets can be independent (or unpaired) if the subjects in each group are different (monadic cells). Conversely, datasets are paired if the same subjects are included in both groups (sequential monadic cells) (Hinkle et al. 1990; Cohen 1992; Stone et al. 1997). All common statistical softwares[2] calculate a *p*-value based on the parameters that are defined (paired or independent samples, sample sizes equal or not, sample variances equal or not). If it is below the chosen threshold for statistical significance (usually 0.10, 0.05 or 0.01) then the null hypothesis is rejected in favour of the alternative hypothesis which is the population means are statistically different with a given level of confidence (90% for a *p*-value of 0.10, 95% for a *p*-value of 0.05 and 99% for a *p*-value of 0.01). In consumer large-scale studies, it is usual practice to highlight significant differences at 95% confidence level. Results at a 90% confidence level can be looked into and interpreted as 'trends' that may need further attention (it is important in that case to be cautious, as

---

1 Alpha and Beta risks: Alpha risk is the risk of incorrectly deciding to reject the null hypothesis. For example, with a 95% confidence level, there is a 5% chance that products are determined different when they are actually not. Alpha risk is also called Type I error or more commonly 'false positive'. Beta risk is the risk that the decision is made that the products are not different when they actually are. Beta risk is also called type II error or 'false negative'. It is calculated based on the power (probability of correctly rejecting the null hypothesis): Power $= 1 - \beta$

H_0: null hypothesis, H_a: alternative hypothesis		Decision	
		Reject H_0	Fail to reject H_0
Actual	H_0 true	Type I α-Risk False positive	**Correct decision** **Confidence** **interval $= 1 - \alpha$**
	H_a true	**Correct decision** **Power $= 1 - \beta$**	Type II β-Risk False negative

2 Microsoft Excel, SAS, XLStats (https://www.xlstat.com/en/), Matlab (https://www.mathworks.com/), SPSS (https://www.ibm.com/us-en/marketplace/spss-statistics).

90% confidence level does not mean there is only a 10% risk of making the wrong decision due to β risk explained in footnote 1).

For illustration purposes, it can be interesting to represent distributions per product. Figure 3.9 shows an example where visually the distribution of liking scores, on a 0-10 scale, shows that sample MD may be more favourably perceived than sample WF.

Structured scales can also be transposed into intensity scales (low end starting by 0 or 1), represented in the same way with bar charts and analyzed with t-test statistics:

I like it very much	=5
I liked it	=4
I neither liked it nor disliked it	=3
I disliked it	=2
I disliked it a lot	=1

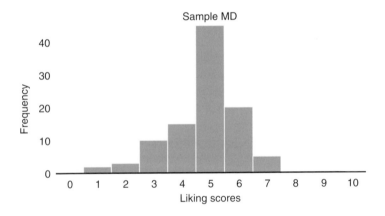

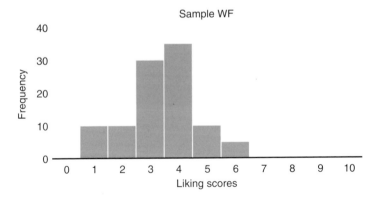

Figure 3.9 Frequency of liking scores for samples MD and WF.

Table 3.7 Top/bottom nomenclature for structured semantic scales.

I like it very much	Very satisfied	Top 1	Top 2
I liked it	Satisfied		
I neither liked it nor disliked it	Neutral		
I disliked it	Dissatisfied		Bottom 2
I disliked it a lot	Very dissatisfied	Bottom 1	

Attributes rated on structured semantic scales are often represented by showing how many respondents chose any given answer or category (value and percentage). Then when several products are compared (monadic cells or sequential monadic), statistics are run to highlight significant differences on these distributions.

In summary charts, top and bottom categories are often aggregated as shown in Table 3.7.

Intensity scales may also be aggregated into top/bottom categories (Tables 3.8 and 3.9).

Once responses are clustered into Top 1, Top 2, … categories, results are commonly presented as shown in Table 3.10.

Table 3.8 Top/bottom nomenclature for intensity scales 1–10.

10	Top	Top 2
9		
8		
7		
6		
5		
4		
3		
2		Bottom 2
1	Bottom	

Table 3.9 Top/bottom nomenclature for intensity Likert scales 1–5.

5	Top	Top 2
4		
3		
2		Bottom 2
1	Bottom	

Table 3.10 Example of large-scale study summary chart (Top 2 boxes).

	% Satisfaction (Top 2)			Mean scores (Top 2)			
	Product A	Product B	Difference	Product A	Product B	Difference	p-Value[a]
Overall satisfaction	81	78	3	5.4	5.3	0.1	NS
Product features							
Ease of usage overall	88	89	−1	5.9	5.9	0	NS
Ease of opening packaging	83	87	−4	5.2	5.8	−0.6	**p < 0.001**
Ease of dispensing	82	72	**10**	5.4	5.4	0.3	**p < 0.001**
Consistency	76	60	**16**	5.3	5	0.3	**p < 0.001**
…	…	…	…	…	…	…	…
Results obtained							
Quality of result overall	92	93	−1	6.2	6.3	−0.1	**p < 0.001**
…	…	…	…	…	…	…	…

[a] 95% confidence level.

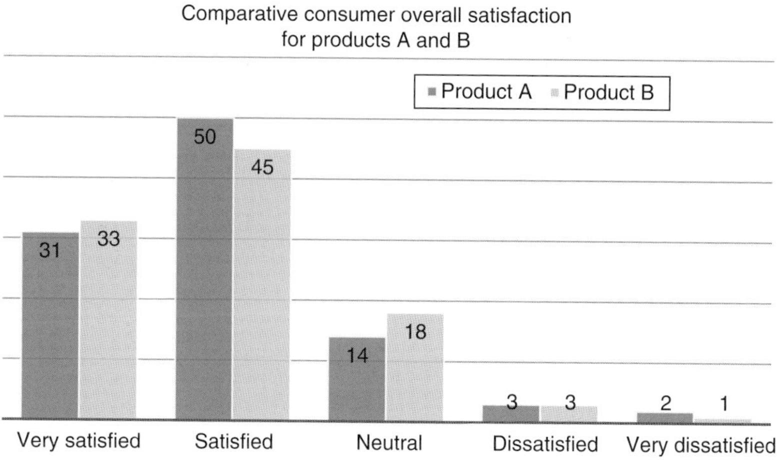

Figure 3.10 Consumer overall satisfaction bar graphs.

In the example of Table 3.10, for each attribute, a p-value below 0.05 indicates that products A and B are significantly different at a 95% confidence level regarding their Top 2 results. For key attributes such as overall satisfaction, comparative distributions of responses may bring an interesting visual representation as shown in the example of Figure 3.10. In this example, it corroborates that these two products do not differ in a major way in terms of overall satisfaction.

Bottom boxes can be represented as well and bring an important perspective in some cases. The example shown in Table 3.11 shows the attribute consistency. For that attribute, Table 3.10 showed Top 2 numbers lower than for the rest of the attributes. It seems obvious that both products may not be entirely satisfying for that attribute, and it is also clear that product B has lower ratings. However, looking at the bottom two numbers further highlights a relative dissatisfaction for product B which is an important aspect to consider for further improvements.

To further understand the different components of consumer satisfactions and the ways to measure and represent it, one can refer to the very comprehensive manual written by Oliver (2010).

Data collected through agreement scale or re-use/re-purchase scales are usually also represented with response distribution charts and top/bottom tables with p-value indicators.

Data collected on JAR 5-point scales is usually represented with bar charts that show the percentages of the JAR levels for the different attributes for each evaluated product (Figure 3.11).

Then, the data is usually analyzed to understand the acceptability scaling (Lawless & Heymann 2010) and identify directions for further product improvement. The method used is commonly called 'penalty analysis'

Table 3.11 Example of large-scale study summary chart (Bottom 2 boxes).

| | % Satisfaction (Bottom 2) | | | Mean scores (Bottom 2) | | | |
	Product A	Product B	Difference	Product A	Product B	Difference	p-Value
Overall satisfaction	5	4	1	1.3	1.2	0.1	NS
Product features							
...	...	...	...	...	...	...	...
Consistency	5	15	-10	1	0.5	0.5	**p < 0.001**
...	...	...	...	...	...	...	...

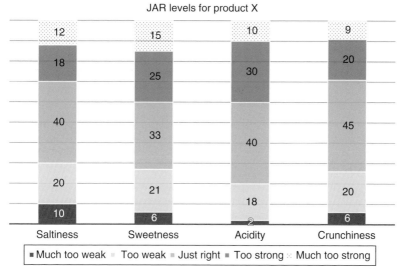

Figure 3.11 Percentages for JAR levels, example for four attributes.

(Ares et al. 2014). On the JAR scale, 'much too weak' (or 'not enough at all') corresponds to 1, 'too weak' corresponds to 2, 'JAR' corresponds to 3, 'too strong' corresponds to 4 and 'much too strong' corresponds to 5. The penalty analysis, based on multiple comparisons through analysis of variance (ANOVA), consists in identifying, for each level of the JAR scale if it is related to significantly different results in the liking scores.

Many statistical softwares offer the modules to run penalty analysis. The various outputs are typically:

- A correlation matrix between liking scores and different JAR attributes (Table 3.12) which shows variables that may be impacting liking scores as well as variables that are correlated or anti-correlated to each other.

Table 3.12 Example of correlation matrix liking × JAR attributes.

	Liking	Saltiness	Sweetness	Acidity	Crunchiness
Liking	1	**0.268**[a]	0.029	−0.021	**0.308**[a]
Saltiness	**0.268**[a]	1	−0.112	**−0.172**[a]	0.100
Sweetness	0.029	−0.112	1	−0.046	−0.033
Acidity	−0.021	**−0.172**[a]	−0.046	1	0.079
Crunchiness	**0.308**[a]	0.100	−0.033	0.079	1

[a] Indicates variables that may have a strong impact on liking scores (95% confidence level).

Table 3.13 Example of penalty analysis results.

	Level	%	Liking scores	Mean drops	p-Value	Penalties	p-Value
Saltiness	Not enough	39.33	4.831	2.187	**p < 0.0001**		
	JAR	37.33	7.018			1.816	**p < 0.0001**
	Too much	23.33	5.829	1.189	**0.018**		

- Penalty analysis outputs which show for each attribute if not enough or too much levels penalize significantly the overall liking of the product. Table 3.13 shows an example where not enough saltiness has a significant negative impact on overall liking scores as it strongly penalizes the products. Too salty has a negative impact as well. In the case of other variables, there may be no significant impact. Both results are very valuable to understand which attributes need the special attention of the product developers (as they drive most of the liking scores) and what aspects need further improvements on a product or not.

Regarding the no-difference/no-preference responses, there are different schools of thought as to what to do. The different alternatives are:

- To drop the data
- To equally and proportionally distribute that data between the different alternatives.

It is always possible to do the analysis with both scenarios and identify the most conservative (the one showing less significant differences) and the most tolerant options.

Lastly, data from ranking scales results into a ranking scores matrix (Table 3.14). Significance of the differences is usually analyzed using the non-parametric Friedman test (Friedman 1940). Some softwares offer alternative test options for this type of data (Wilcoxon 1945; Zimmerman & Zumbo 1993).

Looking at the numbers in Table 3.14, it looks obvious that products A (always ranked 1) and E (mostly ranked 5) are most likely different. However, products B, C and D seem closer visually. A Friedman test will allow to determine whether there are significant differences or not.

Demographic data always needs to be displayed in detail in a report or its appendix, either through data tables (Table 3.15) or via frequency charts.

Under certain circumstances, the consumer base encompassed in a large-scale study can be divided into sub-groups of individuals who have common characteristics, often from demographic or geographic point of view. This can

Table 3.14 Example of ranking results.

Product	A	B	C	D	E
1	1	4	2	3	5
2	1	2	2	3	5
3	1	2	4	3	5
4	1	2	4	4	5
5	1	3	4	4	5
6	1	3	3	4	5
7	1	3	4	2	5
8	1	4	2	3	4
9	1	2	4	2	5
10	1	3	3	2	5
...	...	...	...	...	...
Sum of ranks					

Table 3.15 Demographic information (relative distributions).

Demography	Product A	Product B
Gender		
Male	63	61
Female	37	39
Age		
18–34	2	3
35–44	7	6
45–54	23	24
55–64	44	43
65+	24	24
Household income		
...	...	...

allow to confirm or invalidate hypothesis explaining how products are perceived. A **segmentation** can also be done *a posteriori* based on the outcome of a study like drivers or behaviours expressed by consumers. The sub-groups can be defined based on results on certain attributes that clearly show separated segments. Statistical tools such as agglomerate hierarchical clustering (AHC) can be used to discern existing groups within the data collected (Drake et al. 2009; Giraud et al. 2013). This can only be done if sub-group populations remain large enough to use quantitative statistics (typically with at least $n = 100$). If, prior to running a study, it is anticipated that a segmentation might be needed, it is obviously recommended to recruit a base large enough overall and for the anticipated sub-groups. Cariou and Wilderjans (2017) published recently a study describing a multi-attribute consumer segmentation that can be used as a reference. Identifying sub-groups with specific needs and expectations is key for marketers to further develop brands and concepts that cater those needs (Dibb & Simkin 1991).

3.2.7   Budget Considerations

Costs associated with large-scale quantitative studies include:

- Product samples and labels (may require a pilot plant or large-scale purchase)
- Screener development
- Recruitment of participants, follow-up contacts
- Product placement (if applicable), may require rental of a venue or mileage to reach location where participants may pick up the products
- Participants' incentives
- Questionnaire development and coding into software
- Statistical analysis
- Top-line and report

Most consumer goods companies outsource large-scale quantitative studies as the logistics imply heavy resources. Here again, some Universities or Business Schools may offer this type of services with more cost-efficient offers. However, it is less common than for qualitative projects to find teams that have the appropriate resources.

Otherwise, market-research companies offering large scale capabilities may be found in the same directories mentioned in Section 3.1.6: Blue Book published by The Marketing Research Association (https://bluebook. insightsassociation.org), Green Book published by the New York Marketing Association (https://www.greenbook.org/aboutus). The American Marketing Association (https://www.ama.org) also yearly publishes the AMA Gold report which lists top Market-Research Firms. Most of the top companies run very large-scale studies and operate worldwide with large capabilities to conduct studies locally or in multiple regions or countries.

As for qualitative research, it is important to consider the criteria below:

- Geography: it is often better to choose smaller but locally implanted firms for their knowledge of local habits and language. They also usually offer more competitive prices. However, for large multi-region or multi-country studies, it is often recommended for the same company to oversee projects in their entirety to ensure consistency in questionnaires, analysis and final synthesis. In those cases, it is very important to check questionnaire translations and local adaptations to reflect cultural particularities. For example, different Spanish-speaking countries may not have the same vocabulary to define certain things, different countries or even regions within a same country may have different distribution channels, different predominant brands.

- Knowledge of the industry: some firms are more specialized or have more experience in certain product categories.
- A single long-term partner may be an advantage at times to aggregate data and knowledge over time (in the same format), and therefore look at multiple studies or pool data for a higher statistical power.

Comparative analysis of multiple bids is always recommended before moving forward (see Section 6.3).

Lastly, some studies may be done very cost efficiently using tools such as SurveyMonkey (https://www.surveymonkey.com) or Typeform (https://www.typeform.com). In the case of informational surveys this type of tools offers wide capabilities to create, conduct and analyze surveys.

Lately crowdsourcing practices have also emerged (Brabham 2008), which are applied to the market-research domain. Indeed, some crowdsourcing tools can be used to collect large amounts of information on some products, services or ideas. One of the most commonly mentioned tools is *Amazon's Mechanical Turk* which has the potential of collecting rapidly large amount of user data at extremely low costs. However, questions or tasks need to be short, focused and well formulated to ensure reliability of the information that is collected (Kittur et al. 2008). Some sorting/cleaning of the data collected may also be necessary. Many warn against the lack of assurances that comes with the collected data (Goodchild & Glennon 2010). However, in some specific situations it may be worth trying it, given the extremely low cost involved (Leimeister et al. 2009).

3.3 Ethnographic Studies: In-Depth Exploration of Consumer Needs and Expectations

3.3.1 When to Conduct an Ethnographic In-Depth Study

Ethnographic research stems from qualitative methods previously described. However, its purpose goes far beyond. It also comes largely from the field of anthropology and sociology as it seeks to embrace and study an entire culture in a holistic way. It is used to understand context, document behaviours and usages, frustrations or pains and identify unmet needs. Usually, the focus is on a specific demographic group and a geographic location. More largely it focuses on a specific group or organization in their 'natural habitat' to dive into lifestyle and values. The idea is ultimately to lead to unearthing opportunities. Here again, methodologies can be tailored to fit the brief and deliver the expected insight. By definition, ethnographic approaches consist in the observation of consumer's behaviour in the field (in a non-CLT location), namely a natural setting where consumers are going to be more spontaneously themselves (Atkinson & Delamont 2001). The 'watching' portion is more important than the 'asking' one (Elliott & Elliott 2003).

3.3.2   Define the Market and Sample

The target audience in this type of methodology is always very narrowly defined based on the objective: spending time with teenage girls preparing and shopping for a sweet 16 event, shopping with stay-home moms at local grocery stores, cooking dinner with families, discussing their fridge content, shadowing baby-boomer women during their skin care night routines or active working men during their morning rituals.

Most ethnographies involve around 10–15 participants per segment and per delimited geographic zone. For example, if the objective is to understand millennial Hispanic women make-up routines in the United States, it can be decided to expand the study to three or four geographic locations: Northeast, West Coast, Central and South. In the latter example, it might be necessary to further define origins in terms of countries or skin tones. Depending on product categories, less geographically diverse countries may not require as many sub-divisions provided minimal climate and culture differences have been identified.

3.3.3   Define the Test Design

Here again, as described in Section 3.1.2, defining the study brief is key to design and plan the study to obtain the results that fulfil the defined objectives.

Most of the time, ethnographic studies involve home visits that can last anywhere from 1.30 to 2–3 h. During those visits, a discussion and observation is engaged with the participant in their real-life environment surrounded by the products they have and the reasons to use them. Visits can include observation (and eventual video recording) of products 'usage. Ethnographic studies also encompass 'shop-along' methods, where consumers' shopping behaviours are observed in-store. The idea is to follow a consumer to understand his/her reactions to product's merchandising. It allows to analyze actual shopping attitudes rather than claimed behaviours. It also allows to confront real challenges faced by consumers such as opinions on packaging, labels, shelve placements, availability and price. Both at-home and in store, usually consumers get highly involved in the study provided the rapport that is established is natural and unbiased and often unforeseen reasons or relationships may be discovered.

The approaches that can be used to enter consumers' real-life are almost infinite nowadays beyond traditional immersion as described above. To be less invasive (one of the drawbacks of the immersion can be that some people may modify their behaviour), consumers can be asked to take in-home pictures and/or videos, those can be then discussed more in-depth with probing questions. Discussions can happen remotely with webcams or via smartphones either live (online discussions) or through live bulletin boards. A significant advantage of online tools is that it drastically reduces geographic limitations and often also cost. It is possible to reach out to anyone anywhere theoretically. Consumers can also be requested to fill traditional journals, those can be written or more commonly now digital.

Discussions guides can be conceived in a similar way to those described in Section 3.1.3 for qualitative studies as these are of the same nature. However, the unrolling of the encounter needs to be as spontaneous and non-directed as possible. Hence, the discussion guide will be used as a support to ensure that all briefed topics are covered in a way or another, but the interviewer will primarily adopt an observation and listening role.

Appendix 3.3.3 gives an example of discussion guide for a home visit.

This role adopted by the interviewer, rather invasive and exposed into the consumer's intimate life can lead to some sensitive scenarios. It is always important to anticipate ways of exiting a situation that may be uncomfortable without raising ethical issues. Let us imagine a situation where the interviewer arrives into the targeted address and feels the environment is unsafe. It is always possible, even after crossing the door, to excuse oneself having to take a personal call or having to deal with an emergency; or in another situation where the respondent that opens the door does not correspond to the screener specifications. Rather than detailing arguments that can be misinterpreted, it is always reasonable to, once again, indicate the need to handle a personal situation and offer to call back later.

3.3.4   Define a Timeline

The consumer immersion exercise requires to thoroughly define the context not only geographically but also timewise. The observation needs to be done at the time of the day that makes more sense from the consumer routine standpoint. If the routine that is going to be observed may be impacted by climate or season, the moment of the year needs to be chosen cautiously as well.

An ethnographic study can be relatively extended over time as shown by Figure 3.12.

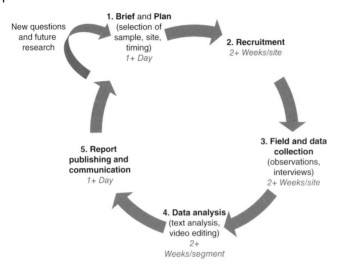

Figure 3.12 Ethnographic research timeline.

3.3.5   Analysis and Deliverables

Ethnographic research reports are typically very visual. They include:

- Pictures (of environment and context, products, gestures, expressions)
- Videos (of gestures during product usage or application, of the interview to capture tone and emotions)
- Deep interview content analysis illustrated with representative and relevant verbatim
- Gestures or product usage can be displayed in very visual and detailed timelines, for example, that illustrates the 'product journey' or routines (Figure 3.13).

3.3.6   Budget Considerations

Budget for ethnographies is usually high and includes, similarly to qualitative studies,:

- Screener development
- Recruitment of participants
- Participants' incentives
- Discussion guide development
- Mileage for home visits
- Interviewer(s) time
- Full analysis and report
- Video recording of observations

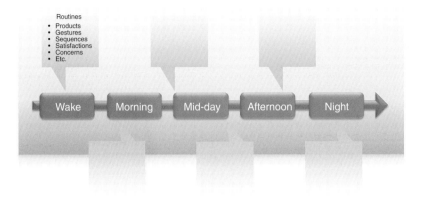

Figure 3.13 Example of daily routine template for report.

Ethnographic studies need to be conducted by specialized companies, usually small and local. Deep knowledge of local nuances and specificities is key. Many freelance researchers offer their service often more cost efficiently than larger firms. Market-research directories mentioned in Section 3.1.6 can be used for reference.

As for all types of studies, it is important to get several quotes before moving forward (see Section 6.3).

3.4   Additional Approaches to Detect Breakthrough Innovations: How to Assess the 'Wow' Factors?

3.4.1   Less Conventional Methods

3.4.1.1   Kano

Kano analysis is a method developed by Professor Kano of the Rika University in Tokyo (Kano et al. 1984; Shenk 2014). It is a quantitative technique that is used to identify features (for products or services) that have a significant impact on desirability and consumer liking. It usually leads to highlight existing identified characteristics that can be enhanced or new dimensions that can be further explored (Sauerwein et al. 1996). It is also particularly interesting for new product or service categories for which there is no a priori knowledge.

For each attribute, in the questionnaire, a question is phrased in two ways:

1) The attribute is present
2) The attribute is not present

Table 3.16 Example of question in a Kano questionnaire.

Functional form	If the helmet allows adjustable ventilation, how do you feel?	1) I really like it that way
		2) I expect it that way
		3) I am neutral
		4) I dislike it but I can live with it that way
		5) I dislike it and cannot accept it that way
Dysfunctional form	If the helmet does not allow adjustable ventilation, how do you feel?	1) I really like it that way
		2) I expect it that way
		3) I am neutral
		4) I dislike it but I can live with it that way
		5) I dislike it and cannot accept it that way

And for each of these two ways, five categories are proposed to the respondent: I really like it that way/I expect it that way/I am neutral/I dislike it but I can live with it that way/I dislike it and cannot accept it that way (Table 3.16).

In the end, all attributes end up being classified into four categories:

1) **Must-be requirements** (also called satisfying **basic needs**): the product suffers extremely if these attributes are absent, to a point that consumer will not be interested in the product at all (typically a clean room in a hotel). However, the consumer takes these for granted or as prerequisites and does not explicitly demand them. They allow a company to get into the market.
2) **Core requirements** (also called '**one-dimensional**' or satisfying **performance needs**): the product 'can never have enough of them', consumer satisfaction is proportional to the level of fulfilment (typically ease of use). They allow a company to remain into the market.
3) **Attractive requirements** (also called satisfying **excitement needs**): the attribute, not always explicitly expressed nor expected, can have the greatest influence on consumer's satisfaction. It adds to the appeal; however, it does not detract if absent (it can be having access to for instance). They allow a company to excel in the market.
4) **To be explored** (or '**questionable**'): attribute does not raise strong positive or negative feelings. It can be because it is irrelevant or because it is not understood but may be worth further assessment before being discarded.

Classifying attributes that way offers the advantage of setting priorities for product development and deciding on which features it is most important to invest.

To start a Kano analysis, it is necessary to first determine a comprehensive list of product features/requirements via a qualitative explorative research (typically focus groups). This needs to be conducted in such a way that it allows to deep dive into consumer problems. Questions to explore are the following:

- What are the associations the consumer makes with the product or product category?
- What are the problems that the consumer associates with the product or product category?
- What are the criteria that the consumer has in mind when purchasing the product?
- What are the improvements that the consumer would like to better fulfil his/her needs?

Once the attributes are listed, the Kano questionnaire is built as shown in Table 3.16. It is important to be very careful in confirming that questions are correctly understood. Combined answers (functional and dysfunctional form) for each attribute are then logged in a table as shown in Table 3.17.

The interpretation of this chart means that, if combining, both answers yield to:

- A: The feature is attractive to the consumer's point of view.
- M: The feature is a prerequisite.
- C: The more the product exhibits the feature the more the consumer will be satisfied.
- I: The consumer is indifferent to the feature, he/she does not care whether the feature is present or not and is not willing to spend more on the feature.
- Q: The question was most likely phrased incorrectly or was misunderstood.
- R: The feature is not wanted and the consumer even expects the reverse.

Table 3.17 Kano log table.

		Dysfunctional				
		1. Like	2. Must-be	3. Neutral	4. Live with	5. Dislike
Functional	1) Like	Q	A	A	A	O
	2) Must-be	R	I	I	I	M
	3) Neutral	R	I	I	I	M
	4) Live with	R	I	I	I	M
	5) Dislike	R	R	R	R	Q

A, attractive; C, core; I, indifferent; M, Must-be; Q, questionable; R, reverse.

Table 3.18 Example of Kano frequency table.

Feature	M	C	A	R	Q	I	Majority
Attribute A	5	34.3	**51.3**	7.5	0	1.8	A
Attribute B	11.4	**46.1**	28.5	10.5	2.2	1.2	O
...	...	...	...	...	...	...	...

Many recommend adding two questions, for each feature, to the questionnaire, as they can further help setting priorities in development plans:

- One to have the consumer indicate the level of satisfaction for the attribute with his/her current product (on a 5- or 7-point scale, for example, from totally unsatisfied to very satisfied).
- One to self-state the importance of the feature (on a 5- or 7-point scale, for example, from completely unimportant to very important).

Most often responses are gathered in a chart that summarizes frequency of answers (Table 3.18).

Cases of very spread results may suggest multiple consumer segments with different expectations. In such cases, results can be a basis for a market segmentation study. Evidently, when interpreting results, there is a hierarchy between attributes in the way they can influence perceived product quality and therefore consumer satisfaction. Results of self-stated importance questions can also be taken into consideration when making decisions towards priorities in product development. Ultimately, it is important to:

1) Fulfil all must-be attributes (M)
2) Gain competitiveness on core attributes (C)
3) Ensure attractive attributes stand out (A)

It is important to remember that consumer needs evolve over time and for some product categories the evolution can be relatively quick. If we take the example of the music industry, it is obvious that between the era of cassettes, then CDs and iPods, needs and expectations have evolved in a very drastic way. Therefore, there is a need to reassess most categories periodically.

Kano method has been used commonly by multiple industries. Some have implemented improvements or new analytical angles. Xu et al. (2009) give an interesting example applied to the design of automotive dashboards.

3.4.1.2 Thurstone Scaling

Thurstone scaling was developed to construct quantitative unidimensional attitude measurements (Sauser 2010). On a given issue, statements on attitudes towards the issue are collected. Those statements typically range from positive to negative, some being more neutral. Then, for each statement, respondents give a rating on how they feel towards those statements. Usually ratings go from 1 (very negative) to 11 (very positive). To illustrate the approach, an example of issue could be 'meat originating from pork raised with antibiotics'. In that example, the statements could be:

- *It should be against the law*
- *There is absolutely nothing wrong with it*
- *It is good to prevent porks from being sick*
- *It should be legalized*
- *It can harm human beings*
- *It can harm children*
- *It is just horrible*
- *It cannot do any harm*

Then ratings for each statement from all respondents are collected and averaged:

It should be against the law	9.1
There is absolutely nothing wrong with it	1.7
It is good to prevent pork from being sick	4.4
It should be legalized	1.5
It can harm human beings	7.6
It can harm children	8.4
It is just horrible	2.0
There is no trace of antibiotics in the meat	4.3
It cannot do any harm	1.3

Lastly, the statements are used in questionnaires, presented to naïve individuals in an ascending order low to high average, ideally representing equal appearing intervals. Individuals are requested to rate them on an agreement scale.

This method leads to identify attitude dimensions that can be measured on defined issues. Some have tried to simplify it (Likert et al. 1993) and its applications have been recently updated and revisited in the domain of consumer choices (Ennis 2016). It ultimately allows to quantify how individuals view (favourably or not) certain topics. However, it is often dismissed as it is considered time consuming and cumbersome. Furthermore, it is not always easy to reach a consensus on some attitudes, especially on neutral ones (Roberts et al. 1999).

3.4.2 Thinking Out of the Box

A large number of relatively well-established methods have been described and for most situations, it is highly recommended to follow guidelines of widely tested approaches to ensure rigour and quality of results. However, it is also important to keep thinking as a researcher despite those rules. A researcher's mind process relies on:

- Always clarifying the research question
- Designing the research proposal based on that question

In some cases, questions may be 'unsearchable' with classical approaches and may require some out-of-the-box more innovative strategies.

Very innovative prototypes may require stepping out of established paths, pilot test new designs, revise and re-test to fine-tune the best way to collect information. Usability of software, of electronic tools, for example, does not always fit into existing checkboxes (Rubin & Chisnell 2008). There are also other examples of situations where it is hard to apply traditional methods. In the case of pet food for instance, the purchase decision is going to be made by the owner. He or she will be sensitive to the presentation and communication around the product on the point of sale primarily or in the advertising. However, the re-purchase will be dependent not only on the owners' impression (does it smell too bad in my kitchen?) but also on the pet's reaction (if the pet eats it quickly and seems happy then it maybe worth it!). Hence, in that category, measure may require more complicate approaches. Some products such as professional hair or make-up products may never be purchased unless recommended by a professional stylist or make-up artist to his/her client. Trust and loyalty towards the professional who advises is very impactful. However, to re-purchase, the client needs to feel satisfied in the long run. Hence, anticipating such channels and behaviours may require more complex testing scenarios.

Over the past decade, some market research companies have been diving into new trails that have been so far little explored with consumer and sensory approaches and require unique logistics and mindsets: testing running bikes, surfing boards or skis, assessing luxury jets, extreme weather gears or even the sound of a lighter (Lageat et al. 2003).

References

Ares, G, Dauber, C, Fernández, E, Giménez, A, Varela, P 2014, Penalty analysis based on CATA questions to identify drivers of liking and directions for product reformulation, *Food Quality and Preference*, vol. 32, issue Part A, pp 65–76.

ASTM 2016, *ASTM E1958-16 Standard Guide for Sensory Claim Substantiation*, ASTM International, West Conshohocken, PA.

Atkinson, P, Delamont, S 2001, *Handbook of Ethnography*, Sage Publications Inc., Thousand Oaks, CA, London and New Delhi.

Aust, L 1998, *Cosmetic Claims Substantiation*, Taylor & Francis and CRC Press, New York.

Belk, R, Fischer, E, Kozinets, R 2013, *Qualitative Consumer and Marketing Research*, Sage Publications Inc., Thousand Oaks, CA, London and New Delhi.

Bernard, H 2000, *Handbook of Methods in Cultural Anthropology*, Altamira Press, Plymouth.

Brabham, D 2008, Crowdsourcing as a model for problem solving, *Convergence: The International Journal of Research into New Media Technologies*, vol. 14, issue 1, pp 75–90.

Brace, I 2008, *Questionnaire Design: How to Plan, Structure and Write Survey Material for Effective Market Research*, Kogan Page limited, London and Philadelphia, PA.

Brannen, J 2005, Mixing methods: the entry of qualitative and quantitative approaches into the research process, *International Journal of Social Research Methodology*, vol. 8, issue 3, pp 173–184.

Bryman, A 1988, *Quantity and Quality in Social Research (Contemporary Social Research)*, Routledge and Taylor & Francis group, London and New York.

Calder, B 1977, Focus groups and the nature of qualitative marketing research, *Journal of Marketing Research*, vol. 14, issue 3, Special Issue, pp 353–364.

Cariou, V, Wilderjans, T 2017, Consumer segmentation in multi-attribute product evaluation by means of non-negatively constrained CLV3W, *Food Quality and Preference*, In Press, Corrected Proof. http://www.sciencedirect.com/science/article/pii/S0950329317300174 (Accessed on August 23, 2017).

Cohen, J 1992, Statistical power analysis, *Current Directions in Psychological Science*, vol. 1, issue 3, pp 98–101.

Desmet, P 2005, Measuring Emotions: Development and Application of an Instrument to Measure Emotional Responses to Products. In: Blythe, MA, Overbeeke K, Monk AF, Wright PC (eds) *Funology*, vol. 3 Human-Computer Interaction, Springer, Dordrecht, pp 111–123.

Dibb, S, Simkin, L 1991, Targeting, segments and positioning, *International Journal of Retail and Distribution Management*, vol. 19, issue 3, pp 4–10.

Drake, S, Lopetcharat, K, Drake, M 2009, Comparison of two methods to explore consumer preferences for cottage cheese, *Journal of Dairy Science*, vol. 92, issue 12, pp 5883–5897.

Edwards, R, Holland, J 2013, *What Is Qualitative Interviewing?* Research Method Series, Bloomsbury Research Methods Series, Bloomsbury, London and New York.

Elliott, R, Elliott N 2003, Using ethnography in strategic consumer research, *Qualitative Market Research: An International Journal*, vol. 6, issue 4, pp 215–223.

Ennis, D 2016, *Thurstonian Models: Categorical Decision Making in the Presence of Noise*, First Edition, The Institute for Perception, Richmond, VA.

Ennis, D, Ennis, J 2016, *Readings in Advertising Claims Substantiation*, First Edition, The Institute for Perception, Richmond, VA.

Fischer, C 2006, *Qualitative Research Methods for Psychologists*, Elsevier, Burlington, MA, San Diego, CA and London.

Fonteyn, M, Kuipers, B, Grobe, S 1993, A description of think aloud method and protocol analysis, *Qualitative Health Research*, vol. 3, issue 4, pp 430–441.

Friedman, M 1940, A comparison of alternative tests of significance for the problem of m rankings, *The Annals of Mathematical Statistics*, vol. 11, issue 1, pp 86–92.

Frith, H 2000, Focusing on sex: using focus groups in sex research, *Sexualities*, vol. 3, issue 3, pp 275–297.

Geary, R 1947, Testing for normality, *Biometrika*, vol. 34, issue 3/4, pp 209–242.

Giraud, G, Amblard, C, Thiel, E, Zaouche-Laniau, M, Stjanović, Ž, Pohar, J, Butigan, R, Cvetković, M, Mugosa, B, Kendrovski, V, Mora, C, Barjolle, D 2013, A cross-cultural segmentation of western Balkan consumers: focus on preferences towards traditional fresh cow cheese, *Journal of the Science of Food and Agriculture*, vol. 93, issue 14, pp 3464–4372.

Goodchild, M, Glennon, J 2010, Crowdsourcing geographic information for disaster response: a research frontier, *International Journal of Digital Earth*, vol. 3, issue 3, pp 231–241.

Hall, A, Rist, R 1999, Integrating multiple qualitative research methods (or avoiding the precariousness of a one-legged-stool), *Psychology and Marketing Special Issue: Qualitative Research*, vol. 16, issue 4, pp 291–304.

Hammersley, M 2013, *What Is a Qualitative Research?* Research Methods Series, Bloomsbury, London and New York.

Hinkle, D, Wiersma, W, Jurs, S 1990, Review: applied statistics for the behavioral sciences, *Journal of Educational Statistics*, vol. 15, issue 1, pp 84–87.

Israel, G 1992, *Determining Sample Size*, Fact Sheet PEOD-6, University of Florida Cooperative Extension Service, Institute of Food and Agriculture Sciences, EDIS, Gainesville, FL, pp 1–5.

Jager, G, Schlich, P, Tijssen, I, Yao, J, Visalli, M, de Graaf, C, Stieger, M 2014, Temporal dominance of emotions: measuring dynamics of food-related emotions during consumption, *Food Quality and Preference*, vol. 37, pp 87–99.

Jessen, S, Kotz, S 2011, The temporal dynamics of processing emotions from vocal, facial, and bodily expressions, *NeuroImage*, vol. 58, issue 2, pp 665–674.

Kano, N, Seraku, N, Takahashi, F, Tsuji, S, 1984 Attractive quality and must-be quality, *The Journal of the Japanese Society for Quality Control*, vol. 14, pp 39–48.

King, S, Meiselman, H 2010, Development of a method to measure consumer emotions associated with foods, *Food Quality and Preference*, vol. 21, issue 2, pp 168–177.

King, S, Meiselman, H, Carr, B 2010, Measuring emotions associated with foods in consumer testing, *Food Quality and Preference*, vol. 21, issue 8, pp 1114–1116.

Kish, L 1963, *Survey Sampling*, Second Edition, John Wiley & Sons, Inc., New York.

Kittur, A, Chi, E, Suh, B 2008, CHI'08 Proceedings of the SIGCHI Conference on Human Factors in Computing Systems, Florence, Italy, April 5–10, 2008, pp 453–456.

Lageat, T, Czellar, S, Laurent, G 2003, Engineering Hedonic Attributes to Generate Perceptions of Luxury: Consumer Perception of an Everyday Sound, Marketing Letters, vol. 4, issue 2, pp 97–109.

Laros, F, Steenkamp, JB 2005, Emotions in consumer behaviour: a hierarchical approach, *Journal of Business Research*, vol. 58, issue 10, pp 1437–1445.

Lawless, HT, Heymann, H 2010, Acceptance Testing. In: *Sensory Evaluation of Food*, Food Science Text Series, Springer, New York, pp 325–347.

Leimeister, J, Huber, M, Bretschneider, U, Krcmar, H 2009, Leveraging crowdsourcing: activation-supporting components for IT-based ideas competition, *Journal of Management Information Systems*, vol. 26, issue 1, pp 197–224.

Lenth, R 2012, Some practical guidelines for effective sample size determination, *The American Statistician*, vol. 55, issue 3, pp 187–193.

Leow, R, Morgan-Short, K 2004, To think aloud or not to think-aloud: the issue of reactivity in SLA research methodology, *Studies in Second Language Acquisition*, vol. 26, issue 1, pp 35–57.

Liamputtong, P 2011, *Focus Group Methodology, Principles and Practice*, Sage Publications Inc., Thousand Oaks, CA, London and New Delhi.

Likert, R, Roslow, S, Murphy, G 1993, A simple and reliable method of scoring the Thurstone attitude scale, *Personnel Psychology*, vol. 46, issue 3, pp 689–690.

Mariampolski, Hy 2001, *Qualitative Market Research*, Sage Publications Inc., Thousand Oaks, CA, London and New Delhi.

McDaniel, C, Gates, R 2005, *Marketing Research*, John Wiley & Sons, Inc., Hoboken, NJ.

McKelvie, S 1978, Graphic rating scales: how many categories? *British Journal of Psychology*, vol. 69, issue 2, pp 185–202.

Miles, S, Frewer, L 2001, Investigating specific concerns about different food hazards, *Food Quality and Preference*, vol. 12, issue 1, pp 47–61.

Oliver, R 2010, *Satisfaction: A Behavioural Perspective on the Consumer*, Routledge, London and New York.

Roberts, J, Laughlin, J, Wedell, D 1999, Validity in the Likert and Thurstone approaches to attitude measurements, *Educational and Psychological Measurement*, vol. 59, issue 2, pp 211–233.

Roininen, K, Arvola, A, Lähteenmäki, L 2006, Exploring consumers' perceptions of local food with two different qualitative techniques: laddering and word association, *Food Quality and Preference*, vol. 17, issue 1–2, pp 20–30.

Rubin, J, Chisnell, D 2008, *Handbook of Usability Testing: How to Plan, Design, and Conduct Effective Tests*, Second Edition, Wiley Publishing Inc., Indianapolis, IN.

Sauerwein, E, Bailom, F, Matzler, K, Hinterhuber, HH 1996, The Kano model: how to delight your consumers, *International Working Seminar on Production Economics*, vol. 1, issue 4, pp 313–327.

Sauser, W 2010, Thurstone Scaling. In: Weiner, IB and Craighead, WE (eds) *Corsini Encyclopedia of Psychology*, Vol. 1, John Wiley & Sons, Inc., Hoboken, NJ.

Schifferstein, H, Fenko, A, Desmet, P, Labbe, D, Martin, N 2013, Influence of packaging design on the dynamics of multisensory and emotional food experience, *Food Quality and Preference*, vol. 27, issue 1, pp 18–25.

Shenk, D 2014, *Kano Analysis*. Wiley StatsRef: Statistics Reference Online.

Smithson, J 2000, Using and analysing focus groups: limitations and possibilities, *International Journal of Social Research Methodology*, vol. 3, issue 2, pp 103–119.

Spiggle, S 1994, Analysis and interpretation of qualitative data in consumer research, *Journal of Consumer Research*, vol. 21, issue 3, pp 491–503.

Stewart, K, Williams, M 2005, Researching online populations: the use of online focus groups for social research, *Qualitative Research*, vol. 5, issue 4, pp 395–416.

Stone, H, Sidel, J, Bloomquist, J 1997, Quantitative Descriptive Statistics. In *Descriptive Sensory Analysis in Practice*, Wiley-Blackwell, London, pp 53–69.

Student 1908, The probable error of a mean, *Biometrika*, vol. 6, issue 1, pp1–25.

Sztainer, D, Story, M, Perry, C, Casey, M 1999, Factors influencing food choices of adolescents: findings from focus group discussions with adolescents, *Journal of American Dietetic Association*, vol. 99, issue 8, pp 929–937.

Van Den Haak, M, De Jong, M, Schellens, P 2010, Retrospective vs. concurrent think-aloud protocols: Testing the usability of an online library catalogue, *Behavior and Information Technology*, vol. 22, issue 5, pp 339–351.

Veludo de Oliveira, T, Ikeda, A, Campomar, M 2006, Laddering in the practice of marketing research: barriers and solutions, *Qualitative Market Research*, vol. 9, issue 3, pp 297–306.

Wilcoxon, F 1945, Individual comparisons by ranking methods, *Biometrics Bulletin*, vol. 1, issue 6, pp 80–83.

Williams, M, Vogt, W 2011, *Innovation in Social Research Methods*, Sage Publications Inc., Thousand Oaks, CA, London and New Delhi.

Xu, Q, Jiao, R, Yang, X, Helander, M 2009, An analytical Kano model for customer need analysis, *Design Studies*, vol. 30, issue 2, pp 87–110.

Zimmerman, D, Zumbo, B 1993, Relative power of the Wilcoxon test, the Friedman test and repeated-measures ANOVA on ranks, *The Journal of Experimental Education*, vol. 62, issue 1, pp 75–86.

4

Study Plans and Strategy: Sustainable Short-, Mid- and Long-Term Vision

4.1 Definition of Key Performance Indicators

A key performance indicator (KPI) is a measurable value that is used to assess how an objective is achieved. Some also called them key success indicators. KPIs are very commonly used as vital management tools in multiple domains such as supply chains (Cai et al. 2009), business, financial and sales perspectives (Marr 2012). In the same way, in our domain, KPIs are to be used linked to the project strategic goals and as tools that allow decision-makers to gauge whether the project is on target towards those goals. In the case of the evaluation of product development projects, forefront KPIs need to be chosen to be 'leading' indicators and give guidance on future. There are also necessary 'lagging' KPIs that will validate solid performance on some metrics but not necessarily forecast future successes. It can take great effort and time to define a relevant set of undisputable KPIs. It is however very important to take that time. It usually requires cross-functional discussions between decision-makers and experts from different functions (marketing, consumer insight, R&D) and fine-tuning them requires diligence from all. The different steps to achieve the consensus include:

- Clarifying the vision and the strategy
- Setting goals and performance measures
- Setting targets (and resources and milestones later on)

Defining KPIs is intimately linked to the objectives and strategy and gives an unbiased tool to focus on when analyzing final outcomes. It allows all parties, managers and stakeholders to assess results objectively and take actions. Of course, for a given project, KPIs need to be revisited regularly as the project evolves.
Key qualities of KPIs are:

- Be well defined and quantifiable
- Be crucial to the goal
- Be clearly understandable and communicable

Consumer and Sensory Evaluation Techniques: How to Sense Successful Products, First Edition. Cecilia Y. Saint-Denis.

In most cases, there are some universal KPIs such as overall liking, overall satisfaction and overall preference measures. Furthermore, additional KPIs are intrinsically linked to a project. Hence, right KPIs for one project might not be the right ones for another one, as they must match the project's specific goals and strategy.

Knowledge held or specifically gathered by consumer insight teams and large-scale U&A studies can be a solid base of information to define KPIs for a project in consensus among all stakeholders. Important needs, expectations, gaps, barriers or frustrations are usually key indicators to be followed when developing a new product to understand how consumer's expectations are met or surpassed. To define KPIs, it is important to first set up a comprehensive list of the product's features and then to establish a hierarchy of priorities. For example, if the subject that matters the most is environmental sustainability in the automotive industry, the top KPIs for the development of a new product could be the drivers of environmental performance such as:

- Fuel consumption/mile
- Average carbon-dioxide emissions

All other features being at least equal.

When not enough is known for the product category, an *ad hoc* thorough research can be conducted to define and prioritize KPIs. For example, performing a KANO study (Section 3.4.1) can be an efficient way to straighten up a comprehensive list of attributes and organize them according to their impact on consumer desirability and liking. Indeed, as previously seen, a KANO study allows to classify requirements in different categories:

- Must-be or basic
- Core performance
- Attractive
- To be explored

It seems reasonable to say that, for must-be and core features, the expectation will be to reach at least parity with the benchmark, whereas for the attractive features superiority will be required to move on. Superiority can also be required for core attributes in some instances, especially if there is a real need or gap identified. Superior results on 'to-be-explored' attributes may be a 'nice-to-have' result to further valorize or expand in the product communication, but those may not be determining factors to move forward or not.

Thus, parity or superiority versus a benchmark, on a defined list of features, is typically the heart of the KPI-based measurement tool. However, in some cases, there is no benchmark, especially when considering a brand new disruptive product. In those cases, for each key feature, it will be necessary to consensually define the threshold to agree on a successful result. It could be decided that to get a 'green light' and move on, the Top 2 box on *overall liking* measure needs to be equal or superior to 85%. If, the goal of the project with the new proposal is to fulfil a specific need not yet covered by existing products, it can be decided to set as a threshold a 90% Top 2 box on satisfaction for that specific attribute. The level of challenge will fully depend on the category that is being entered, the size and value of the market, the risks and opportunities it may open further on. If venturing in a totally new domain, it may be possible to define and prioritize the features to focus on upfront but it may be impossible to define targets or thresholds before having run any test.

Once the KPI tool associated with the project has been established, at each step, a scorecard or dashboard (Kaplan & Norton 1996; Lee & Sai On Ko 2000) is to be established to analyze the test results and recommend steps to follow. It allows to indicate, in a relative straightforward way, whether results allow to get a 'green light' or 'red light'. Mixed results may lead to a 'yellow light' with eventually a decision to move on, having measured controlled risks, or a decision to improve certain aspects before moving on. These may be called key improvement indicators (KIIs) (Setijono & Dahlgaard 2007). Having these clearly identified KIIs can enhance efficiency when going back in the process to reduce those defects or enhance those gaps, as additional tests may only focus on those before moving forward.

Scorecards can have different formats overall and depend on the type of test that is being analyzed.

However, they will always be based on:

- The objectives
- The measures
- The targets
- The results obtained
- The recommendations based on those results and the context

A template that can be used to build scorecards and ultimately devise on a set of strategies stems from Lee and Sai On Ko (2000) 'SWOT analysis' shown in Table 4.1. It allows to cross results (measures vs targets) and market context (opportunities and threats).

Strengths represents the strong points, the unique or distinct advantages that make the product stand out as a breakthrough innovation or versus existing products.

Table 4.1 Template for SWOT analysis.

	Strengths (S)	Weakness (W)
Opportunities (O)		
Threats (T)		

Weaknesses are areas where competition operates better or has an advantage. Therefore, it represents areas that can or need to be improved.

Opportunities show aspects where there is room for growth, emerging markets or trends where the product may show good potential.

Threats are the domains where competition may be encroaching the products' ground, where new trends or societal changes may put at risk the need for the product or service.

From there, balanced score cards, as described by Kaplan and Norton (1996), translate the project's objectives into a set of performance measures organized strategically to facilitate decisions (Table 4.2).

The following sections show examples of applications for different types of tests and contexts. Figure 4.1 summarizes the different steps that are detailed below and that are needed in the innovation process as they determine the

Table 4.2 Template of scorecard.

KPI attributes	Strengths	Weaknesses	Recommendations (based on results and opportunities/threats)
Attractive			
Core performance			
Must-be			
To be explored			

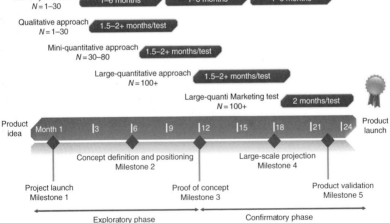

Figure 4.1 Product development strategy.

time-to-market. The exploratory phase represents the "product design" phase after the consumer or market need has been identified and the idea launched. The confirmatory phase represents more specifically the "product development" phase once the concept is validated. Of course, once the product is fully validated, there is going to be a supply chain laps of time to make and distribute the product which will be under Operations responsibility and no longer under Research and Development responsibility. It is important to always keep that in mind in the projection vision.

4.2 Exploratory Phase

The rising cost of tests have lead development teams in the industry to build testing plans along the product development process to be able to eventually 'kill' or 're-orient' a product earlier than later if need be. The goal is to run extensive pre-testing, showing that a product is potentially a 'winner', to ensure that large-scale, more expensive, tests are used as much as possible in a confirmatory way. Indeed, what industries want to avoid is to learn later in the process something that could have been learned earlier, especially if that leads to terminate a project.

A real success is most likely guaranteed if a 'go' is not decided just when a product 'does not look bad' but rather when it 'truly looks good'. For that, it is important to ignore the pressure to get something into the market and stop any product that does not appear to be a potential true winner. To make the right decisions, it is important to come up with the right criteria to judge evaluation results. There is often a conflict between those in a company who want higher levels for earlier kill (usually researchers/developers) and those who prone lower levels to have a larger number of new products as an outcome (usually marketers). Hence, some fine screening criteria need to be established to reconcile both objectives. Thus, KPIs and scorecards described in the previous section need to be used as early as possible in the evaluation process, even during exploratory phases where measures and targets will be harder to assess as they are in most cases not entirely numerical.

In addition to the screening criteria, it is also important to clearly define rules to eventually override them, especially during exploratory phases where some not fully finalized proposals may need second chances if there is a 'gut feeling' or if some interesting strengths and opportunities are overshadowed by some defaults. Thereby, these override situations will necessarily come up. However, they need to be kept as an exception.

Furthermore, in earlier exploratory stages, as well as later in confirmatory stages, it is important to keep in mind that decisions are always made in conjunction with multiple information sources. Indeed, in addition to sensory and consumer results, often other elements need to be considered such as microbiology results that guaranty stability of perishable products and instrumental measures that define additional properties. Thus, examples of decision paths shown below are highly simplified compared to real-life project management and decisions.

4.2.1 Use of Consumer Insight

This is the very first step in a development process: identify micro, macro and mega trends (in the field as well as in other domains such as technological, socio-economic and environmental) and market dynamics, as well as consumer wants and needs not being met, emerging behaviours or even new consumer segments (Figure 4.2). This allows to draw the future short, mid and longer term. Let us say, for example, that an in-depth consumer insight investigation is done on mothers' perception of strengths and weaknesses of available products to treat babies' dry skin. Available data, social media, additional focus groups or in-depth interviews lead to hypothesis that may need to be validated with more quantitative surveys and ultimately lead to strategic opportunities and actionable innovation roadmaps with new product platform initiatives. The number of selected new projects will then be based on resources available, history of the organization, business priorities and global strategy, profit plans and pay-out length estimates (Cooper et al. 2016a, b).

Consumer insight can also be used to test the viability of certain ideas or concepts before even refining a prototype. Understanding how a concept may be received does not have to wait and can be assessed before starting the development work.

Lastly, consumer insight will also provide the developers with some guidance in terms of pricing and distribution channels: where will the product be sold or the service be offered and how much can it cost to be viable?

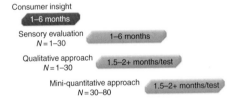

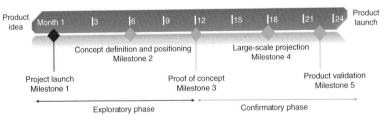

Figure 4.2 Consumer insight and project launch.

4.2.2 Use of Sensory Evaluation

Sensory evaluation provides extremely useful information all along the development project (Figure 4.3). At the earliest stages, it will primarily be used to assess whether first prototypes are on target compared to the initial objectives. Scorecards allow to understand intrinsic strengths and weaknesses of a product versus existing offers on the market or versus a defined benchmark. They can also allow to compare different candidates. Table 4.3 shows an example of a scorecard for a new orange juice formulation tested versus the top selling benchmark on the market on sensory attributes. The example is fictitious for the purpose of the demonstration. But if we imagine that a previous study was done to prioritize attributes, the scorecard can be organized in a way that allows easier decisions.

Based in a scenario such as the example of Table 4.3, it would be recommended to rework the formula to reduce the two aftertaste defaults without modifying the rest of the formula which seems to have multiple strengths, especially on attractive attributes which could represent a competitive advantage. It can be noted that, in this example, the sensory profile of the new formula does not contain segmenting notes such as exotic fruit notes or texture aspects like the presence of pulp. If that is the case, it is necessary to keep in mind that differences between the formulas that are compared may lead to different consumer preferences in different segments.

As shown in Figures 4.4, 4.5, 4.6 and 4.7, sensory evaluation is a decisional tool that can intervene at any moment during the development process, on both exploratory and confirmatory phases. Indeed, data collected via a sensory

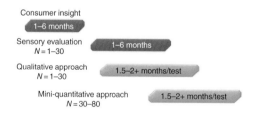

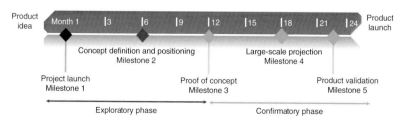

Figure 4.3 Sensory evaluation and concept definition and positioning.

Table 4.3 Scorecard for sensory evaluation results (example of new orange juice development vs market leader benchmark).

KPI attributes	Strengths	Weaknesses	Recommendations (based on results and context)
Attractive	Thickness > ** Caramel note > *** Floral note > ** Pineapple note = ns Lemon peel note = ns		The new formula offers very interesting strengths on most attractive and core attributes. It is at parity for the
Core performance	Orange intensity = ns Mandarin intensity = ns Cooked orange note > **	Bitter aftertaste > ** Metallic aftertaste ≥ *	must-be ones and shows a few areas that can be further explored
Must-be	Sweetness = ns Acidity = ns	Artificial note = ns	In the core domain, there are a couple of defaults that appear higher in the new formula (one is significant, one shows a trend)
To be explored	Astringency < ** Apricot flavour > ** Orange peel flavour > ***		
Final recommendation			

ns, non-significant.
***$p \leq 0.01$; **$0.01 < p \leq 0.05$; *$0.05 < p \leq 0.1$

expert panel is extremely robust in describing the intrinsic characteristics of a product objectively, provided the panel has been trained and checked correctly as described in Chapter 2.

4.2.3 Use of a Qualitative Approach

When a prototype seems promising, it is recommended to proceed with a qualitative test with the goal of achieving a proof of concept (POC) (Figure 4.4). This means to give the product to a small group of consumers to get early feedback on the product perceived performance. The data that is collected, as seen in Section 3.1, is verbal content that is analyzed as such. The product can be tested alone or in comparison with a reference or a benchmark, both being presented blind in a sequential monadic design.

Table 4.4 shows another fictitious example of a new orange juice formula that is tested versus a top selling benchmark on a group of 15 consumers.

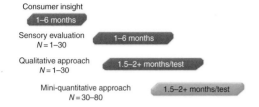

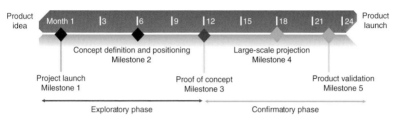

Figure 4.4 Qualitative approach and concept definition and positioning.

The objective of this new formula is to find a proposal that has a competitive advantage versus the market leader in a very saturated market (threat). The decision to have the consumers also test the benchmark blind allows to understand how it is perceived without its marketing cover and to have a more solid base to compare the number of citations and make decisions. If we refer to the Table 3.4, going from 'a few' to 'half' or from 'half' to 'almost all' may indicate aspects to focus on. The alternative can be to recruit 15 people who are 'heavy' or 'exclusive' consumers[1] of the benchmark and take the benchmark as a reference in a portion of the interview. However, when possible, it is preferable to retest the benchmark blind to have a clean objective read. Indeed, some consumers do not have a clear memory of their usual product. Furthermore, some products in the market context may be very well perceived and once tested in blind receive bad ratings. It is interesting to grasp that in the decision process as some intrinsic advantages can be better valorized later on.

Sometimes it is not possible to easily recruit exclusive or heavy users of certain brands or products in very volatile markets. Indeed, in some product categories, consumers are not always loyal to a brand, either because the offer is very large or because they purchase primarily based on promotions and coupons. Also, some consumers tend to have multiple brands simultaneously in their households to just be able to vary and alternate. In a similar way, for very innovative or even disruptive products, there might be no benchmark to use as a reference. In those cases, interviews are going to purely focus on the

---

1 Rules for the screener can be defined based on usage frequency. Screening could be done based on filters such as 'has exclusively used the brand over the past 6 months', or be more flexible: 'the last used product is of that brand' or 'the brand represents the "go to" product'.

Table 4.4 Example of scorecard for qualitative results.

KPI attributes	Strengths n/15 mentions for test (n/15 for bench)		Weaknesses		Recommendations (based on results and context)
Attractive	Juice is thicker	13/15 (2/15)			The new juice formula is described in positive terms for its thickness by almost all. Over half mention positively caramel-like and floral notes
	Caramel, brown sugar, syrup	11/15 (7/15)			
	Floral notes	9/15 (4/15)			
	Lemon mentions	3/15 (3/15)			
Core performance	Appropriate orange taste	14/15 (13/15)	Overripe taste	2/15 (0/15)	In regard to core attributes, half the panel mentions a pleasant cooked/marinated orange taste
	Clementine taste	12/15 (11/15)			
	Cooked/marinated orange taste	7/15 (0/15)			
Must-be	Appropriate sweetness	15/15 (14/15)	Artificial note	1/15 (2/15)	Compared to the benchmark, half mention a less rough feeling in mouth and over half feel there are other fruit flavours which are perceived positively
	Appropriate acidity	13/15 (14/15)			
To be explored	Smooth (not rough) in mouth	8/15 (2/15)			It is important to note a that two consumers feel an unpleasant overripe taste
	Various fruits mentioned	9/15 (5/15)			
Final recommendation					

description of the tested product and the sentiment attached to the way it is perceived. Numbers in the scorecard are interpreted overall with common sense: on a panel of 15 consumers for instance, above-half mentions are usually an encouraging result. Different approaches are possible. Most often, it is recommended to run different tests to address different potential targets. For example, if a product is developed to address an unmet need, it will be naturally important to recruit those who have expressed that unmet need. In some cases, the product may be tested among consumers who have abandoned the category because of that unmet need or because of a deep dissatisfaction. In case of a totally new offer, consumer insight may give guidance on different segments to address. In all those cases, scorecards can be more 'tolerant' in the sense that a lower number of citations may still be encouraging as a premise of a new territory.

Based on the scenario of Table 4.4, it would most likely be recommended to move on to the next step. The couple mentions of an 'overripe taste' may not be considered alarming, if the researcher knows that they have a link with the 'cooked/marinated' citations and that overall for those consumers it was far from crippling but rather anecdotic. It is always important to know the context: dive into the interviews' transcripts if necessary, but also assess threats and opportunities from the bigger picture. Of course, in a real-life context, it is unlikely that this scorecard would be a standalone result. Any decisions need to be made looking at the global picture of other simultaneous qualitative tests, sensory tests, as well as any other supporting data that the organization has.

4.2.4 Use of a Mini-Quantitative Approach

A green light after a qualitative approach allows POC and usually leads to a mini-quantitative test (Figure 4.5). Many organizations opt for this approach as it is less expensive and involving than a large quantitative test. Skipping it does not necessarily mean a potential failure on the large quantitative approach. However, the risk may be higher without that additional screening step. In cases where the qualitative test gives extremely promising results, it may be decided to go straight to a large quantitative test, especially when product development is held by very tight deadlines or limited budget. The one path that is conversely not advisable at all is to run a mini-quantitative test in lieu of a large quantitative test to save time and money. Indeed, a mini-quanti, even though based on a larger sample size ($N = 30–80$) than the qualitative remains largely exploratory and needs to be interpreted with much caution. Projections of mini-quanti results to the real population are risky. *A fortiori* even more if the results are mixed. Hence, a mini-quanti step brings additional assurance but may be skipped if tight deadlines impose drastic choices in favour of a quantitative test.

Scorecards to summarize results from a mini-quanti can be presented in the same way as those for qualitative results showing number of responses

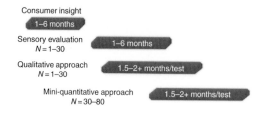

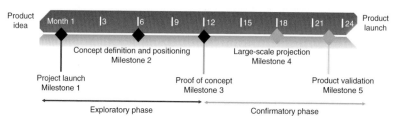

Figure 4.5 Mini-quantitative approach and proof of concept.

(especially when sample sizes are on the lower side of $N = 30$). The reports usually focus on graphs showing response distributions for each item. It is also possible to show frequencies with Top 2/Bottom 2 boxes, p-values and significances (Table 4.5) when sample size is on the larger side. Indeed, mini-quantis are for the most part constituted with close-ended questions. Therefore, it is possible to run statistical tests and p-values. However, those are always to be interpreted with much caution. A significant p-value on a mini-quantitative test definitely indicates a trend to be further investigated. However, extrapolation to a larger population may not necessarily be accurate.

Here again, it is recommended, when possible, to test versus a reference or a benchmark following a sequential monadic design. However, as for the qualitative tests, sometimes it is not feasible. For instance, if the new product is very innovative or, if for logistic reasons, sourcing enough samples of the benchmark is challenging or too expensive. In the case of a monadic test, it will be necessary to determine beforehand consensually the acceptable thresholds for the different items (what numbers do we judge acceptable for the Top 2 boxes of key attributes for instance). For very established products, the frame of reference could be existing test results in the same category. For very disruptive products, it may be necessary to be less conservative and allow lower standards.

The fictitious example shown in Table 4.5 shows mixed results. Given the threat that represents a saturated market and a very well-established bench, not obtaining winning global indicators is a strong handicap despite the attractive and core performance strengths. The bitterness may also be an area of improvement. Overall, even though there is an opportunity to penetrate

Table 4.5 Example of scorecard for mini-quantitative results.

KPI attributes	Strengths (Top 2 boxes)	Weaknesses (Bottom 2 boxes)	Recommendations (based on results and context)
Attractive	Juice consistency[a] > 85**		The new juice formula seems to exhibit attractive and core performance strengths compared to the benchmark: its consistency in the cup, as well as its texture in mouth that is particularly appreciated. Its overall flavour and the uniqueness of its fruit notes. However, slightly high cumulated Bottom 2 boxes for bitterness in comparison to the bench may require special attention. Global indicators (overall liking and satisfaction) do not indicate an overall winning formula. Overall scores are high for both formulas
	Uniqueness of fruit notes > 76**		
Core performance	Overall liking ≥ 88*	Bitterness > 15**	
	Overall satisfaction ≥ 90*		
	Overall flavour > 87**		
	Orange taste = 91 ns		
	Clementine taste = 77 ns		
Must-be	Sweetness = 89 ns	Artificial = 25 ns	
	Acidity = 76 ns		
	Colour = 95 ns		
To be explored	Texture in mouth > 86**		
Final recommendation			

ns, non-significant.
[a] Original questions are phrased as the following example: *How satisfied are you with the juice consistency? Very dissatisfied/dissatisfied/not dissatisfied not satisfied/satisfied/very satisfied.* Top 2 boxes cumulates responses for satisfied and very satisfied.
**$0.01 < p \leq 0.05$; *$0.05 < p \leq 0.1$

the market with a proposal that seems attractive, it is necessary to strengthen the formula, either intrinsically or via its positioning and story. The table indicates the Top 2 box numbers for the test product and whether it is significantly different from the benchmark. A more detailed presentation may also include Top 2 values for the benchmark as a reference. However, it may lead to a scorecard more difficult to read.

4.3 Confirmatory Phase

4.3.1 Use of a Quantitative Approach

A large quantitative test is intended to confirm or invalidate previous positive results (Figure 4.6). A product can be tested as a standalone in a monadic cell design or versus a reference in a sequential monadic design. Two products may also be compared by each being tested on parallel identical monadic cells. If a product is tested as a standalone (most often because there is no identified benchmark or reference to compare it to), in the same way as for mini-quantitative tests, it will be necessary to determine beforehand consensually the acceptable thresholds for the different items. For product categories established in the market, previous quantitative test results could be used as a frame of reference of numbers that are to be attained. For very disruptive products, where there is no frame of reference, agreeable thresholds need to be established consensually by the project management team.

Table 4.6 shows an extract of how a scorecard for a large quantitative test comparing two products (either two monadic cells or a sequential monadic) can be presented (in a true scenario the scorecards are typically much more extensive). KPIs (attributes) are to be organized according to agreed priorities and categories. In most cases, cumulated Top 2 boxes are good indicators. It can also be decided to build the same scorecard with Top 1 boxes to be more conservative. Bottom 1 or Bottom 2 boxes can be displayed as a means to focus on weaknesses. If there are only a few, they can be incorporated in one unique chart.

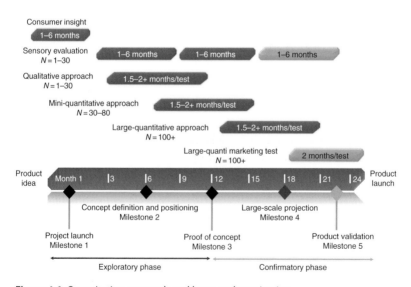

Figure 4.6 Quantitative approach and large-scale projection.

Table 4.6 Example (extract) of scorecard for a large quantitative.

KPI attributes		Juice A (test)	Juice B (ref)
Attractive	Juice consistency	85**	80
	Uniqueness of fruit notes	76**	71
Core performance	Overall liking	88*	86
	Overall satisfaction	90*	89
	Overall flavour	87**	79
	Orange taste	91 ns	92
	Clementine taste	77 ns	79
	Bitterness	20** (Bottom 2)	10 (Bottom 2)
Must-be	Sweetness	89 ns	91
	Acidity	76 ns	75
	Colour	95 ns	94
To be explored	Texture in mouth	86**	82

Recommendation (based on results and context)

Few attractive, core and nice-to-have attributes with scores higher for the test versus the reference

Overall liking and satisfaction not significantly higher

ns, non-significant.
**$0.01 < p \leq 0.05$; *$0.05 < p \leq 0.1$

Table 4.6 shows a scenario where there are a few attractive, core and nice-to-have attributes where scores are higher for the test versus the reference. However, the global indicators, overall liking and satisfaction are not truly significantly higher than the reference. At the level of a quantitative test, those are always under high scrutiny. Here, they are just higher at a 90% confidence level. On a very competitive market, that may not be considered as enough to move forward. It would be of course a management decision. It is important to keep in mind, that it is very common to obtain very encouraging results in the beginning of a development process (qualitative stages) and as the 'vise tightens' differences may narrow as well. It is a quite classic scenario. This does not prevent at times extremely positive results at a quantitative stage, which may raise great enthusiasm.

4.3.2 Product Validation

Marketing test is often considered the last check before rolling the product out into real-life distribution (Figure 4.7). Many consider the marketing test stage as a 'dress rehearsal' of all the product elements together as its intent is to

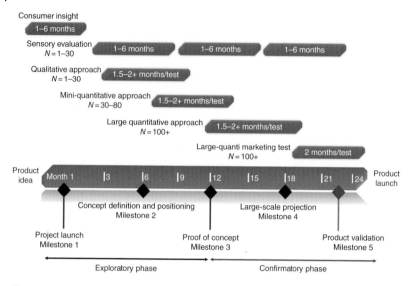

Figure 4.7 Marketing test and product validation.

anticipate real-world through a representative microcosm. The truth is, often when a marketing test is launched, the decision to go has already been made and in the absence of truly bad results, the plan continues. However, results can alter the plan by giving a no-go, in which case, the marketing test becomes a disaster check. The disaster check can be on different levels, first one of course being consumer acceptance, whereas others can be any product aspect that can go wrong on a large scale that is better to know before a launch (packaging problems, stability issues, label issues, and even adverse reactions from consumers such as allergies). Conversely, a successful marketing test will not only reinsure everybody on chance of success but it can also be a tool for sales forces to obtain acceptance and distribution.

Marketing tests can be done following different scenarios:

- Test solely advertising/concept
- Test product samples
- Test both

In all cases, scorecards can be organized in a very similar way as those shown for a large quantitative test on Table 4.6.

In the end, all along this decision chain, it is important to keep in mind that the process needs to be rigorous but also keep some room for intuition. It is also important to always consider what is at stake. Indeed, launching a new frozen cooked dish does not imply the same risk as launching a new car model. It is important to find a good balance between never moving forward if a test does not fully open the go/no-go gate and moving without any kind of validation test.

4.3.3   R&D and Marketing Intertwined Roles

Some new products are initiated by developers as a result of their creativity or of new technologies. Some new products are developed based on an identified consumer need. Both push and pull processes feed the society with new offers. Roles of Research and Development and Marketing teams are often intertwined as research will create a product and marketing will 'dress it up':

• Develop a concept, a story
• Select a name
• Design or fine-tune a packaging
• Develop a communication strategy

Effective communication between R&D and Marketing teams all along a new product development process is a key factor for success (Gupta et al. 1986; Ernst et al. 2010). Indeed, converting ideas into tangible products involves expertise from both sides. Unfortunately, it may often be problematic (Lovelace et al. 2001). The importance of that communication and ways to improve it has been well documented over the past decades (Souder 1988; Pinto & Pinto 1990). However, tensions may be enhanced by a more and more challenging market place.

Conflicts between the two entities will always be inherent to the at times antinomical points of view. However, in order to favour 'productive' conflicts rather than dysfunctional ones it is important to foster:

• Frequent communication
• Bi-directional information exchanges
• Accuracy, transparency and quality of the content exchanged

4.4   Necessary Reconsiderations and Back and Forth

Ultimately the product is ready to be launched following an efficient decision chain (Figure 4.1). However, the path to that launch, and hopefully success, is not always a non-stop, one way smooth trip. Often a yellow or red light imposes a reverse path for reformulation and adds time to what is initially anticipated (Figure 4.8). It is important to keep that uncertainty in mind when planning and forecasting (Cooper 2000; Derbyshire & Giovannetti 2017). Indeed, initially solidly planned scenarios are subject to, not only the inherent risks of the project (not so good test results) but also to socio-economic areas and any kind of disruptive events, new market challenges and even changes in organization priorities.

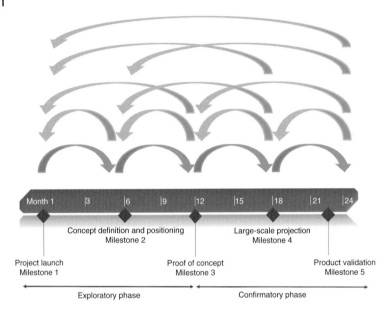

Figure 4.8 Bi-directional multi-paths to success.

4.5 Spin-Offs to Capitalize on Successful Products

True innovations are difficult to achieve. Therefore, when one is obtained, it is important to enjoy the momentum and make the most of it from its 'spin-offs'. Nowadays, with the market becoming more and more saturated and challenging, the paradigm of 'closed innovation' is shifting towards the new one of 'open-innovation' (Chesbrough 2006). Boundaries become less rigid and best innovations can come from the exterior of the company: other industries, universities or start-ups. In that same spirit, when a new technology is built into a winning product in one particular category, in one department or one area of the company, it can be expanded into additional SKUs or even transferred and transformed into new product categories (Jiao et al. 2007). Also, a winning product in a region or country may be tried in another part of the world.

If the same, or slightly modified formula is tried in another region or country of the world, the plan can be deployed right away, starting with a qualitative research locally and moving onto mini-quantis and quantitative large scale when green lights are obtained. In theory, sensory tests provide universal intrinsic product characterisations that can be transposed anywhere. However, that is not always true. Indeed, usages may vary from one region to another. Therefore, a sensory protocol used somewhere may not be inappropriate somewhere else. This is particularly true for personal care products for which,

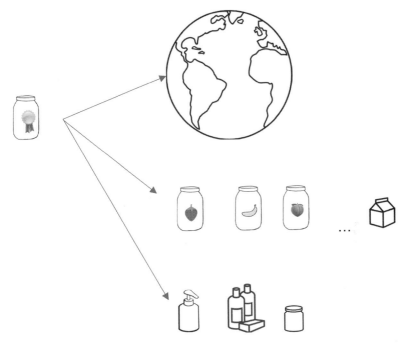

Figure 4.9 Spin-offs of winning product.

in addition, skin or hair properties may impose the necessity to develop local sensory panels to characterize product properties in a relevant way.

If a technology is transposed into a new product category, then again, the decision chain can be re-initiated to define the concept and prove it (Figure 4.9).

References

Cai, J, Liu, X, Xiao, Z, Liu, J 2009, Improving supply chain performance management: a systematic approach to analysing iterative KPI accomplishment, *Decision Support System*, vol. 46, issue 2, pp 512–521.

Chesbrough, H 2006, *Open Innovation: The New Imperative for Creating and Profiting from Technology*, Harvard Business Press, Boston, MA.

Cooper, L 2000, Strategic marketing planning for radically new products, *Journal of Marketing*, vol. 64, issue 1, pp 1–16.

Cooper, R, Edgett, S, Kleinschmidt, E 2016a, Portfolio management in new product development: lessons from the leaders – I, *Research Technology Management*, vol. 40, issue 5, pp 16–28.

Cooper, R, Edgett, S, Kleinschmidt, E 2016b, Optimizing the stage-gate process: what best-practice companies do – II, *Research Technology Management*, vol. 45, issue 6, pp 43–49.

Derbyshire, J, Giovannetti, E 2017, Understanding the failure to understand new product development failures: mitigating the uncertainty associated with innovating new products by combining scenario planning and forecasting, *Technological Forecasting and Social Change*, In Press, Corrected Proof. http://www.sciencedirect.com/science/article/pii/S0040162516302980 (Accessed on August 23, 2017).

Ernst, H, Hoyer, W, Rübsaamen, C 2010, Sales, marketing and research-and-development cooperation across new product development stages: implications for success, *Journal of Marketing*, vol. 74, issue 5, pp 80–92.

Gupta, A, Raj, S, Wilemon, D 1986, A model for studying R&D-marketing interface in the product innovation process, *Journal of Marketing*, vol. 50, issue 2, pp 7–17.

Jiao, J, Simpson, T, Siddique, Z 2007, Product family design and platform-based product development: a state-of-the-art review, *Journal of Intelligent Manufacturing*, vol. 18, issue 1, pp 5–29.

Kaplan, R, Norton, D 1996, *The Balanced Scorecards: Translating Strategy into Action*, Harvard Business School Press, Boston, MA.

Lee, S, Sai On Ko, A 2000, Building balanced scorecards with SWOT analysis, and implementing "Sun Tzu's The Art of Business Management Strategies" on QFD methodology, *Managerial Auditing Journal*, vol. 15, issue 1/2, pp 68–76.

Lovelace, K, Shapiro, D, Weingart, L 2001, Maximizing cross-functional new product teams' innovativeness and constraint adherence: a conflict communications perspective, *Academy of Management*, vol. 44, issue 4, pp 779–793.

Marr, B 2012, *Key Performance Indicators (KPI): The 75 Measures Every Manager Needs to Know*, Pearson Financial Times Publishing, Harlow and New York.

Pinto, M, Pinto J 1990, Project team communication and cross-functional cooperation in new program development, *Journal of Product Innovation Management*, vol. 7, issue 3, pp 200–212.

Setijono, D, Dahlgaard, J 2007, Customer value as a key performance indicator (KPI) and a key Improvement indicator (KII), *Measuring Business Intelligence*, vol. 11, issue 2, pp 44–61.

Souder, W 1988, Managing relations between R&D and marketing in new product development products, *Journal of Product Innovation Management*, vol. 5, issue 1, pp 6–19.

5

Real-Life Anticipation with Market Factors: Brand, Concept, Market Channel, Price

5.1 Highly Challenging Markets

According to Nielsen, 8650 FMCG new products were launched in 2015, totalizing 24,500 new SKUs across Western Europe. Among these, only 18 were classified as breakthrough innovation winners.[1] Examples of those winners launched in 2015 are: Ariel 3-in-1 pods' detergent, Volvic juicy beverage, Scholl velvet Smooth Express Pedi footcare and Garnier Ultimate blends. Beyond statistics on real breakthrough innovations, it may also sound scary to realize that, according to Nielsen again, 85% of the new CPG products fail in the market place. When coming from small companies, these failures may often be due to weak financial resources implemented. When coming from bigger companies, it is often due to a wrong positioning or to a poor differentiation from the existing offer on the market. Tests performed during the development phases can be incriminated if the brief was poorly designed or if there were errors in the data collection or analysis. However, other variables may also impact how well a product performs that may not be fully predicted by studies previously described:

- The nature and context of the market being considered
- The initial brand positioning and strategy on that market if already implanted
- The type of new product (refinement or breakthrough)
- The communication elements around the new product
- The launch context (strong competitive pressure imposing tight deadlines, industrial challenges to achieve the manufacturing of the new product)

---

1 For Nielsen, a breakthrough innovation is defined as meeting three key criteria: deliver a new proposition—not just refinement, generate at least 10 million Euro sales in the first year of trading and maintain at least 90% of their sales in the second year. However, breakthrough or disruptive innovation may have several definitions, not necessarily linked to outstanding sales and economic success, but game-changing/life-changing contribution opening ground-breaking territories for products, services or experiences.

Consumer and Sensory Evaluation Techniques: How to Sense Successful Products,
First Edition. Cecilia Y. Saint-Denis.
© 2018 John Wiley & Sons Ltd. Published 2018 by John Wiley & Sons Ltd.

- How well the product as a whole is in sync with consumers' expectations. Intrinsic qualities need to confirm what the consumer is expecting due to communication; otherwise deception may result in rejection. Furthermore, even if the need has been identified, it is important that the product is introduced at the right time, not too early nor too late and in a way that the consumer understands it. A little bit like a satellite that targets the geostationary orbit, the 'firing window' may be very precise.

5.2   Blind Versus Identified Quantitative Tests

As stated in Section 1.2.1.2, a test product should be blinded when the purpose is to assess or compare product formulations. When a blinded product is presented, no brand expectation exists. Hence, the intrinsic performance qualities of the formulations may be magnified. Thus, blind tests should be primarily conducted during early stages of product development or restages. Some of the questions that a blind test tries to answer are:

- What is the optimal formula?
- What attributes drive consumer linking?
- How does this new (restaged or less expensive) formula compare to the previous one or to the competitor?

However, it is important to keep in mind that under blind conditions, usually differences may be overstated. Conversely, if differences are not measured in a blind test, it is quite unlikely that difference may be measured in a real-life situation.

Consequently, it is important to follow blind tests by **identified tests** where consumers are exposed to the products in 'real-life' situations that bring expectations closer to their daily life. This could entail disclosing the **brand**. In addition, exposure to a **concept** is appropriate when the objective is to understand if the product delivers based on the positioning.

It is most often recommended to conduct identified tests in single monadic cells, under natural settings (typically home) and allowing the consumers to have enough exposure to the product so they can use it repeatedly especially for newer products. Indeed, some products may raise enthusiasm in the beginning and wear out on the long run. Also, often the impact of a brand or a label information may be lessened at some point when consumers taste or truly experience the products (Levin & Gaeth 1988).

Of course, in identified tests scenarios, there is always a halo effect of the brand that needs to be considered. Popular brands will usually lead to higher scores as opposed to less-known ones. To better understand this effect, it is usually recommended to recruit balanced numbers of users of the brand in question and users of competitor brands. Thereby, if a new product is being assessed, it is possible to understand how it appeals to new users and how it can allow to recruit current users of a given brand. However, overall, in identified tests, it is always difficult to understand attribute effects on overall liking independently from the brand effect as most of the time brands have the deepest influence (Chang & Tseng 2015).

5.3 Specificity of Concept Tests

Many products do not find their place on the market because they do not have a strong enough concept. Ultimately the products are not visible enough and rotations become insufficient for the products to be viable. Testing concepts quantitatively may be long and expensive, not to mention the creation of truly engaging concepts which can be very challenging as well. Many companies first proceed with internal screening phases (involving R&D, marketing, operations), then move on to enlarged screening with design and advertising agencies, up to running small- and large-scale studies with consumers. Qualitative approaches are important as they allow to collect early reactions to concepts in the making. Some use these qualitative approaches, in the form of workshops for example, with internal or external participants, in an iterative way during the creation process.

Major steps to assess a concept test are:

Present the concept:

- ✓ A statement or verbal text
- ✓ Graphic visuals (sketch, photos)
- ✓ Videos
- ✓ Audio material
- ✓ Prototypes sample

Measure overall concept reaction:

- ✓ Concept appeal, acceptability, desirability, interest
- ✓ Concept need, what it brings or improves versus existing
- ✓ Likelihood to use or purchase

Example: *Which statement best describes how much you think you would like or dislike this product?*

Like extremely	Like very much	Like quite well	Like somewhat	Like not very much	Not like at all
☐	☐	☐	☐	☐	☐

Conduct a more detailed analysis of concept reaction:

✓ If the concept is made of different components or elements, breakdown of likes and dislikes. Each attribute or component can be evaluated
✓ Assess novelty and superiority versus existing products on the market for each component

Example: *How would you rate the 'natural ingredients' component in terms of being new and different from other products currently available in this category?*

☐ Extremely new and different
☐ Very new and different
☐ Somewhat new and different
☐ Slightly new and different
☐ Not at all new and different

Understand potential usage based on concept:

✓ Likelihood of usage
✓ Identify usage situations
✓ Identify frequency of usage
✓ Compare to similar existing products

Example: *How frequently would you use this product?*

More than once a day	Once a day	2–3 times a week	Weekly	Every 2–3 weeks	Monthly	Every 2–3 months	Once or twice a year	Never
☐	☐	☐	☐	☐	☐	☐	☐	☐

Analyze perceived value:

✓ Estimate product value range
✓ Identify the purchase channel that matches the concept

Example: *Where do you think you would be more likely to find this product?*

☐ Mass merchandiser (Target, Walmart, etc.)
☐ Drug store (CVS, Walmart, etc.)
☐ Grocery store
☐ Warehouse store (Costco, Sam's, etc.)
☐ Internet

Demographics:

✓ Understand market segments that are likely to be attracted

(Questions such as, *who do you think is more likely to be interested in this product?*, with a multiple-choice answer option, could be directly asked. Or segments could be assessed by cross-referencing demographic questions asked in the beginning of the survey.)

Some concepts may be tested *in situ* either in real stores, in reconstituted stores (in a CLT) or even virtual stores (on computer systems now available with high-resolution images, where consumers can almost feel immersed and with possibilities to introduce new products, images, packaging virtually). Some companies now use recent techniques such as eye tracking methods (Duchowski 2007; Wedel & Pieters 2008; Holmqvist et al. 2011) to understand what first attracts consumers' eyes, how long their vision focuses on certain aspects and how quick they lose interest. Reconstituted or virtual stores may also be used to observe and analyze purchase behaviours, in general, with similar approaches as ethnographic studies previously described but with a marketing perspective (Belk 2006).

Semiotic theories and methods have been applied lately in marketing communication as a means to understand how certain symbolic elements (logos, cultural symbols, icons, texts) resonate with target audiences. Semiotics studies the 'signifier' (what is being represented) in relation to what is being 'signified' (the message that is effectively communicated) usually without much words. Hence, semiotics seeks to understand how consumer attitudes and behaviours are formed in relation to popular culture and trends and how marketing can best reach out to consumers in their needs with an accurate and improved communication (Copley 2004; Daymon & Holloway 2011).

5.4 Notions of Modellization

The way certain product components (ingredients, instrumental measures, sensorial aspects, packaging components or even price, brand) are linked to consumer acceptability (liking, satisfaction, preference) is often worth being investigated and determined as it can be a strong tool to guide developers for present improvements or for future outgrowth. It is also a key asset for marketers and sellers to position a new offer and develop the right data-driven communication. Hence, it is a very frequent approach to establish statistical models between these different datasets. It is important to be aware of dangers to which these modellizations may lead and how to prevent them with a meticulous validation.

Modellization always entails finding systematic links between two or more datasets; the following nomenclature is usually used:

P = number of products (1, 2, 3, ... p).

X = explanatory variables (1, 2, 3, ... x). x_{ip} designates the ith explanatory variable for product #p.

Y = dependent variables (1, 2, 3, ... y). y_{ip} designates the jth dependent variable for product #p.

For example, we could have a scenario with eight different vanilla puddings, 10 sensory variables and two hedonic variables (Figure 5.1).

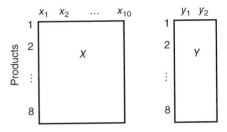

The model in Figure 5.1 serves to understand the nomenclature typically used but is very simplified as one important rule to study the links between two datasets is that the information they cover must be as comprehensive as possible. Indeed, explanatory links that are found may be biased or completely false if the reason for some data relies in another phenomenon that is not included. Another important aspect to verify before running a modellization is to ensure that the data is of good quality:

Figure 5.1 Modellizing impact of 10 sensory variables on two consumer measures for eight products.

- accuracy, repeatability of measurements (for sensory and instrumental)
- representativeness of the sampling (for consumers)

Lastly, it is a requisite to make sure a relation between the two or more datasets exists before proceeding to a modellization. To ensure that, some 'symmetric methods' such as multiple factor analysis (MFA) can be used (Escofier 1994; Pagès 2005; Bécue-Bertaut & Pagès 2008). These are descriptive methods that do not need explanatory and dependent datasets to be defined. The datasets also need to contain enough data for the model to be run.

Once all these prerequisites are fulfilled, the statistical modellization can be run to obtain a representation of the phenomenon, called regression, that links the datasets in a form of an equation that laises each dependent variable y_{ip} with the explanatory variable x_{ip}. The equation obtained becomes a prediction tool. However, extrapolations are always risky as the model is only valid for the domain covered by the datasets and there are no such things as universal models.

Models obtained classically contain the three following elements:

- The function f of the relation between dependent and explanatory variables. That relation could be linear or non-linear. The choice of that function needs assumptions on the nature of the phenomenon often guided by the nature of the data.
- The parameters or weights a_i that specify the relation between dependent and explanatory variables.
- The error of the model ε.

$y_i = f(x_i) = a_0 + a_1 x_1 + a_2 x_2 + \ldots + a_i x_i + \varepsilon$ (in case of a linear model, other functions could be x_i^2, $\log(x_i)$, etc.)

Most softwares allow the selection of multiple models and also the validation of the final ones which is a fundamental step to ensure no bad serious consequences due to wrong modellizations. Main points to verify are:

- Ensure the error ε, which is the unexplained portion of the model, is totally random and encompasses no structure. Otherwise, that would mean it contains explanatory elements that have not been considered. Most softwares offer a graphical representation of errors to check that aspect.
- Calculate the prediction error for the model. This is typically done using cross-validation techniques (the model is calculated without one variable of the dataset and that variable is then predicted and used to calculate the prediction error; in an iterative way, all variables are used to calculate the global error.

Ultimately a validation needs to be done with data that is not encompassed in the model. When very large sample sizes are available, those can be split into two so that the model is built with half of the data and that the validation is done with the other half.

5.5   Preference Mapping and Its Variants

The main objective of any sensory and consumer research is to uncover which are the elements that drive consumer acceptability. Indeed, the purpose is always to optimize consumer response. Hence, data modellizations are developed to understand how descriptive and analytical data (that describes the product characteristics and quality) can explain consumer choices. It is commonly admitted that a measure such as 'overall liking' can be driven by many dimensions. Indeed, two products can have the same rating on an overall liking scale but be liked or disliked for different combinations of reasons. So, in a multidimensional product space, there can be several ideal points.

There are multiple methods that have been published to identify drivers of liking by modellizing product characteristics as explanatory variables (sensorial, instrumental) versus consumer acceptance as dependent variables (usually based on overall liking, but it can be other indicators such as overall satisfaction). Most allow to obtain both the statistical models and the graphical visual representations:

- External preference mapping originated by Schlich and McEwans (1992), a few examples in English can be found in: Ares et al. (2010), Allgeyer et al. (2010), Neely et al. (2010), Ares et al. (2011)
- Its variant, the internal preference mapping (Rousseau et al. 2012)

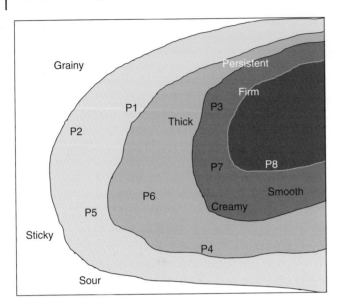

Figure 5.2 Schematized example of a preference mapping.

Internal preference mapping consists in running a PCA (see Section 2.3.1.5) on the consumer dataset and then project the sensory descriptors as supplementary variables on the map obtained. External preference mapping consists in first running a PCA on the sensory data and then running a regression on sensory principal components to explain consumer liking. The model of the regression can be linear or non-linear and ultimately allows to get the weights and function of each original sensory variable on the consumer-liking variable. External preference mapping is one of the most used models as it also allows visual representations that are easy to interpret and communicate. Figure 5.2 shows zones of equal acceptance (different shades of grey, the darker the grey the more appreciated), how products are positioned in terms of appreciation and how sensory attributes are linked to lower or higher appreciation zones.

- Partial least square regression (Bayarri et al. 2011; Ares et al. 2014; Morais et al. 2014).

Partial least square is a common alternative that allows to run a regression on multiple dependent consumer variables at a time. Additional methods have been published giving interesting results:

- Unfolding (Busing et al. 2010)
- Comparison with consumer preferences (Delgado et al. 2013; Predieri et al. 2013)
- Landscape segmentation analysis (Thompson et al. 2004; Rousseau et al. 2012)
- Ideal profile method (Worch et al. 2014).

Some teams have compared multiple methods and introduced penalty analysis in the mix (Lovely & Meullenet 2009). A good reference manual that gathers recent published papers in this domain was edited by Ennis et al. (2014). One can also refer to the very comprehensive manual on the different multivariate and probabilistic methods that can be used in this domain (Meullenet et al. 2007).

5.6 Incorporation of Market Factors in Modellizations

Costa and Jongen (2006) clearly state how important it is to put the consumer's mind and angle of view at the forefront of any innovation strategy to avoid the risk of failing. Searching the optimal product needs to be done not only from a formulation standpoint, but also from a holistic standpoint. Indeed, beyond the product intrinsic characteristics and measures, there can be multiple additional explanatory extrinsic variables like packaging, price as well as global context to explain consumers' arbitration who ultimately find a compromise between quality and price depending on their own personal environment (Figure 5.3). Indeed, being able to predict purchase and re-purchase behaviours is undoubtedly one of the most complex tasks. Nowadays, some mathematical simulations are used to predict market share based on promotional expenditures, awareness, trial and repeat purchase. Tenenhaus et al. (2005) developed the PLS path that allows to encompass multiple explanatory datasets in a model. This method extracts from each explanatory dataset the most influential variables and allows to weight multiple sources among each other. Therefore, this approach allows to understand consumer acceptance taking into account the entire product universe.

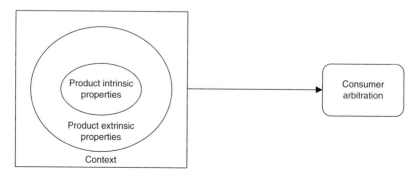

Figure 5.3 Elements influencing consumer arbitration.

References

Allgeyer, L, Miller, M, Lee, S 2010, Drivers of liking for yogurt drinks with prebiotics and probiotics, *Journal of Food Science*, vol. 75, issue 4, pp 212–219.

Ares, G, Giménez, A, Barreiro, C, Gámbaro, A 2010, Use of an open-ended question to identify drivers of liking of milk desserts. Comparison with preference mapping, *Food Quality and Preference*, vol. 21, issue 3, pp 286–294.

Ares, G, Dauber, C, Fernández, E, Giménez, A, Varela, P 2014, Penalty analysis based on CATA questions to identify drivers of liking and directions for product reformulation, *Food Quality and Preference*, vol. 32, issue Part A, pp 65–76.

Ares, G, Varela, P, Rado, G, Giménez, A 2011, Identifying ideal products using three different consumer profiling methodologies. Comparison with external preference mapping, *Food Quality and Preference*, vol. 22, issue 6, pp 581–591.

Bayarri, S, Marti, M, Carbonell, I, Costell, E 2011, Identifying drivers of liking for commercial spreadable cheeses with different fat content, *Journal of Sensory Studies*, vol. 27, issue 1, pp 1–11.

Bécue-Bertaut, M, Pagès, J 2008, Multiple factor analysis and clustering of a mixture of quantitative, categorical and frequency data, *Computational Statistics and Data Analysis*, vol. 52, issue 6, pp 3255–3268.

Belk, R 2006, *Handbook of Qualitative Research Methods in Marketing*, Edward Elgar, Cheltenham.

Busing, F, Heiser, W, Cleaver, G 2010, Restricted unfolding: preference analysis with optimal transformations of preferences and attributes, *Food Quality and Preference*, vol. 21, issue 1, pp 82–92.

Chang, A, Tseng, T 2015, Consumer evaluation in new products: the perspective of situational strength, *European Journal of Marketing*, vol. 49, issue 5–6, pp 806–826.

Copley, P 2004, *Marketing Communications Management: Concepts and Theories, Cases and Practices*, Elsevier Butterworth-Heinemann, Oxford.

Costa, A, Jongen, W 2006, New insights into consumer-led food product development, *Trends in Food Science and Technology*, vol. 17, issue 8, pp 457–465.

Daymon, C, Holloway, I 2011, *Qualitative Research Methods in Public Relations and Marketing Communications*, 2nd Edition, Routledge, London and New York.

Delgado, C, Crisosto, G, Heymann, H, Crisosto, C 2013, Determining the primary drivers of liking to predict consumers' acceptance of fresh nectarines and peaches, *Journal of Food Science*, vol. 78, issue 4, pp 605–614.

Duchowski, A 2007, *Eye Tracking Methodology, Theory and practice*, Springer-Verlag, London.

Ennis, D, Rousseau, B, Ennis J 2014, *Tools and Applications of Sensory and Consumer Science*, The Institute for Perception, Richmond, VA.

Escofier, B 1994, Multiple factor analysis (AFMULT package), *Computational Statistics and Data Analysis*, vol. 18, issue 1, pp 121–140.

Holmqvist, K, Nyström, M, Andersson, R, Dewhurst, R, Jarodzka, H, Van De Weijer, J 2011, *Eye Tracking: A Comprehensive Guide to Methods and Measures*, Oxford University Press, Oxford.

Levin, I, Gaeth, G 1988, How consumers are affected by the framing of attribute information before and after consuming the product, *Journal of Consumer Research*, vol. 15, issue 3, pp 374–378.

Lovely, C, Meullenet, J 2009, Comparison of preference mapping techniques for the optimization of strawberry yogurt, *Journal of Sensory Studies*, vol. 24, issue 4, pp 457–478.

Meullenet, J, Xiong, R, Findlay, C 2007, *Multivariate and Probabilistic Analyses of Sensory Science Problems*, Blackwell Publishing, Ames, IA.

Morais, E, Cruz, A, Faria, J, Bolini, H 2014, Prebiotic gluten-free bread: sensory profiling and drivers of liking, *LWT: Food Science and Technology*, vol. 55, issue 1, pp 248–254.

Neely, E, Lee, Y, Lee, S 2010, Drivers of liking for soy-based Indian-style extruded snack foods determined by U.S. and Indian consumers, *Journal of Food Science*, vol. 75, issue 6, pp 292–299.

Pagès, J 2005, Collection and analysis of perceived product inter-distances using multiple factor analysis: application to the study of 10 white wines from the Loire Valley, *Food Quality and Preference*, vol. 16, issue 7, pp 642–649.

Predieri, S, Medoro, C, Magli, M, Gatti, E, Rotondi, A, 2013, Virgin olive oil sensory properties: comparing trained panel evaluation and consumer preferences, *Food Research International*, vol. 54, issue 2, pp 2091–2094.

Rousseau, B, Ennis, D, Rossi, F 2012, Internal preference mapping and the issue of satiety, *Food Quality and Preference*, vol. 24, issue 1, pp 67–74.

Schlich, P, McEwan, J 1992, Preference mapping: a statistical tool for the food industry. *Science des Aliments*, vol. 12, issue 3, pp 339–355.

Tenenhaus, M, Vinzi, V, Chatelin, YM, Lauro, C 2005, PLS path modelling, *Computational Statistics and Data Analysis*, vol. 48, issue 1, pp 159–205.

Thompson, J, Drake, M, Lopetcharat, K, Yates, M 2004, Preference mapping of commercial chocolate milks, *Journal of Food Science*, vol. 69, issue 9, pp 406–413.

Wedel, M, Pieters, R 2008, A review of eye-tracking research in marketing, *Review of Marketing Research*, vol. 4, pp 123–147.

Worch, T, Crine, A, Gruel, A, Lê, S 2014, Analysis and validation of the ideal profile method: application to a skin cream study, *Food Quality and Preference*, vol. 32, issue Part A, pp 132–144.

6

Internal Studies Versus Sub-Contracting

6.1   Outsourcing: When and When Not?

Outsourcing may often be motivated by cost-saving reasons on operational aspects or due to insufficient internal resources (Nijssen & Frambach 1998). Ultimately, the main pros of outsourcing a consumer study, whether small or large scale, are:

- It guarantees an objective third party and reduces the risk of biasing results (it is actually often essential, especially for claim tests).
- Professional market-research companies have wide knowledge in the testing field, experienced teams and up-to-date tools for more efficient data collection and analysis that usually guarantees good quality results.
- Professional market-research companies usually have access to large consumer databases that allow a higher volume quality recruitment more cost-efficiently.
- Outsourcing to a market-research company typically allows a quicker turnaround for data analysis and reports. Hence the benefit is to be able to focus available internal resources more strategically on things that add value to the business and its competitiveness. It allows to save time to make intelligent decisions.

Despite compelling benefits, outsourcing implies certain constraints and may have drawbacks in some circumstances. Among them:

- For many companies, outsourcing implies going through a heavy internal purchase process: collecting several bids, selecting one, establishing multiple contracts (on the content of the work to be carried out, on confidentiality), validating each step, following up on field progress, back and forth on outcome before final 'goods' acceptance and following up on multi-step payments.

Consumer and Sensory Evaluation Techniques: How to Sense Successful Products, First Edition. Cecilia Y. Saint-Denis.
© 2018 John Wiley & Sons Ltd. Published 2018 by John Wiley & Sons Ltd.

- In companies driven by innovation where new technologies and intellectual property is key, there can be a concern in working with market-research companies that can also be running projects with competitors. Indeed, without implying unethical practices, the risk of revealing a new product idea to a competitor may be higher when outsourced.
- In some situations, cost of outsourcing may imply very high budgets and it may become worth considering running studies in-house. Nowadays, there are several dozens of online survey platforms and plethora of analytical tools for all types of data (numerical, text, social media), many of them being low-cost or even having free basic modules.[1] Hence, tasks that used to be time-intensive can now often be done much more easily in-house.

However, there are real costs to run tests internally, out of the pocket and staff time, and some can often be underestimated: Survey platform tool, recruitment/panel management tool, analysis tools (statistics analysis software, text analysis tools), human resources such as a project leader who may run the field or a scientist who may analyze the data.

The decision on whether to run tests internally or externally comes to five main key angles:

1) Feasibility
2) Responsiveness
3) Quality
4) Security
5) Costs

Thereby, when deciding to run a study in-house, one good approach is to do a pilot test to 'dry run' resources and logistics aspects and therefore assess those five key dimensions before moving on a larger scale and/or longer term. Then, it is possible to weigh the trade-offs of outsourced versus in-house more accurately.

Many softwares come with try-out periods and it is a very good idea to take advantage of those before investing on a longer term. It is important to assess on real-life scenarios whether the tools are easy to use, whether they have the necessary features, whether they are easy to set-up or customize if needed and whether internal staff has time and skills to run them. Indeed, some tools require specific expertise, like programming skills. Often specialized market-research companies invest in staff that has that type of expertise whereas it may be hard to develop that internally. When assessing free and low-cost tools, the points to consider are:

- Does the tool integrate a panel access?
- Does the tool allow quotas and filters?

---

1 Checkbox: http://www.checkbox.com/
SurveyMonkey: http://www.surveymonkey.com/
SurveyGizmo: http://www.surveygizmo.com/

- Does the tool have a library of common questions?
- Are question displays very simple or are there more complex presentations?
- How does the data export to other softwares such as SPSS?
- Do the tools allow basic or more complex statistical analysis?
- Does the tool offer automatic reports?

It is important to realistically gauge the amount of work and skills needed with existing staff, with additional training, with potential new hires or with potential consultants. Indeed, besides the capacity to handle more work (or to give up some current tasks), certain skills are also required such as:

- Project management skills: following up and maintaining targets on schedule, budget, logistics.
- Reporting: analytical and synthetic skills to produce a straightforward output (including technical skills such as statistics or verbal content analysis).
- Best practices in quality control to deliver free-of-error results.
- Ability to present results in front of an audience using experience and context to enhance credibility. Furthermore, capacity to deliver poor results which may be more challenging in front of executives. For that matter, an outside messenger may be easier.

Ultimately, determining whether outsourcing or not comes down to a cost/benefit analysis. There is a huge benefit in obtaining efficiently good quality information with limited staff time. That may or not outweigh the cost implied. It all comes to the budget available in the company for consumer studies to be compared to the cost of investing on internal resources. Nonetheless, it is important to keep in mind that outsourcing requires internal full-time personnel to coordinate work with the outsourcing company.

Small businesses, that do not have internal resources, may outsource more systematically, larger businesses may do certain studies internally and commission others outside. It is quite common in large businesses to conduct at least part of the small-scale steps internally and outsource most of the large-scale tests. Indeed, most constraining aspects of large-scale studies are resources to recruit a large number of consumers, as well as logistics needed for the field portion.

6.2   Precautions When Outsourcing

Market-research companies have competencies to conduct studies efficiently. However, outsourcing not only means giving up budget but also elements of control. Under those circumstances, a few aspects need to be considered with caution.

To succeed, it is key to have an open, honest and respectful **communication** with outsourcing partners. Having clear priorities, objectives and strategy before approaching the company and sharing them will allow to collectively determine the type of approach that needs to be used. Therefore, it is always better to be specific with the market-research partner on the information that is being looked for. It is also substantial to share, as far as possible, how the data is going to be used. For example, is the information going to directly determine or postpone a launch? Is it going to find a positioning among competitors? When possible, sharing background information may also be helpful in finding common solutions. Disclosing, in some cases, longer partnerships that may be envisioned can help the company adjust its resources and better fulfil the needs on a short and longer term. Lastly, concrete practical elements such as existing questionnaires or expected report templates are also elements that can ease communication and ensure that end results match expectations.

Most established market-research companies will produce **quality** results. However, it cannot be excluded once in a while to come across amateur researchers. Therefore, it is very important to assess the preparedness of the company doing the outsourcing. It is important to gauge their quality standards and eventually share the internal quality standards that need to be followed, and ensure that they are understood and applied. Occasionally, general auditing practices and shadowing the way the field is conducted is highly recommended, especially when working with a company for the first time. On the long run, regularly attending field work and interviews is a good practice to also ensure everything is run in the way it was discussed and agreed, to pick up eventual misunderstandings, rather earlier than later. Indeed, errors and misleading results lead to wrong decisions, waste of time and can have dramatic consequences.

Transparent communication is key; however, beforehand security needs to be ensured via a **Non-disclosure agreement** (NDA) sometimes called confidentiality disclosure agreement (CDA) (Klee 2000). The role of this legal contract established between the two parties is to minimize the risk of leakage. It details the elements that are shared but that are to be restricted to any third parties. Via this contract, the two parties signing agree not to disclose information covered by the contract. NDAs can be unilateral if only one party anticipates disclosing confidential information or bilateral when the two parties involved anticipate disclosing information. An NDA needs to be drafted by a legal department and typically contains:

- the parties involved
- the elements that are confidential
- the period of confidentiality

It is a very common and recommended practice to sign a bilateral NDA between a company and a market-research firm, even before the first encounter. It is important to know that even though the risk is reduced when an NDA is signed, it is not completely abolished. Therefore, very sensitive information may not be shared.

When outsourcing a study to a market-research company, as stated above, it is still necessary to have staff coordinate the **logistics** of the study that is being outsourced. Indeed, to ensure a smooth progress there are multiple steps that require robust coordination between internal and external interlocutors:

- Finalizing the detailed test specifics
- Requesting bids
- Analyzing bids, confirming the study is moving on, preparing and signing contracts
- Preparing and sending samples
- Validating screeners, questionnaires, discussion guides
- Following up on recruitment, discussing difficulties and constraining criteria if needed
- Following up on field, shadowing when appropriate, ensuring there are enough products and no issues.
- Following up on agreed deadlines
- Acknowledging receipt of top-line, report
- Validating final 'goods' reception
- Ensuring payments are done in a timely manner

6.3   Criteria to Select a Market-Research Company for a Specific Study

Nowadays there are different types of market-research companies:

- Companies that conduct standard, classic and quick concept or product tests
- Companies that conduct customized small- and large-scale studies
- Companies that go beyond standard result analysis and recommendations pushing the analysis towards potential estimation and sharper market positioning

Market-research companies can also be generalist or specialized:

- Some are more experts in blind products (more used to work with R&D), identified packaged products, concepts, up to brand and market development and potential assessment (more used to work with marketing).
- Some may be more used to qualitative studies or qualitative approaches.
- Some may be specialized by targets: children, youth, seniors and so forth.
- Some may be specialized by product types: food, cosmetics and so forth.
- Some may have unique specialties such as sensory, sniff-tests and so forth.

Finding the right company can be done using the same process as to find any kind of vendor (Quinn & Hilmer 1994). It is important to use word of mouth, listen to colleagues and professional network connections, check references and past clients if possible. When establishing a first relation, it is important to ask questions and be as specific as possible on business needs to ensure they are met by the company's skills area.

Once contact is established, certain considerations are key:

- Does the company seem to have a sensitivity for the considered industry field and consumer segment?
- Does the company seem to understand the objectives?
- Is the company pro-active in suggesting relevant methodological approaches or just ready to run any kind of study?
- Does the company seem competent for the methods that are going to be put in place? Is the company sharing similar case studies?
- For small-scale studies, is the researcher in charge competent for the analysis of verbal content?
- For large-scale studies, do they have staffed statisticians or do they use consultants? In both cases what are their credentials?
- Does the company show interest and implication in a longer term strategy?
- What is the cost? Is there a clear breakdown?
- What is timeline of the proposal?

Ultimately, it is important to always gather and compare several proposals before deciding. Then, there are aspects that can only be evaluated once work formally starts with the company; those may determine future partnerships:

- Does the company hold agreed deadlines?
- Is the quality of results irreproachable?
- Is the report reader friendly and complete?
- Is their presentation of results convincing?

Market-research directories given in Sections 3.1.6 and 3.2.7 can be used as reference.

References

Klee, M 2000, The importance of having a non-disclosure agreement, *IEEE Engineering in Medicine and Biology Magazine*, vol. 19, issue 3, pp 120–120.
Nijssen, E, Frambach, R 1998, Market research companies and new product development tools, *Journal of Product and Brand Management*, vol. 7, issue 4, pp 305–318.
Quinn, J, Hilmer, F 1994, Strategic outsourcing, *Sloan Management Review*, vol. 35, issue 4, pp 43–56.

Appendix

Chapter 1

Section 1.2.1.2

Example of a Questionnaire for a Blind Product Evaluation Followed by Concept Assessment

<u>Respondent identification</u>
Consumer code:
Period of use:

<u>Overall</u>

Q1 How many times did you use the moisturizing cream over the week?

Less than 5 times   ☐
5 times   ☐
6 times   ☐
7 times   ☐
More than 7 times   ☐

Q2 When did you use the moisturizing cream?

In the morning   ☐
At night   ☐
Both morning and night   ☐

Consumer and Sensory Evaluation Techniques: How to Sense Successful Products,
First Edition. Cecilia Y. Saint-Denis.
© 2018 John Wiley & Sons Ltd. Published 2018 by John Wiley & Sons Ltd.

Q3 Overall, how satisfied are you with this moisturizing cream?
Very satisfied ☐
Satisfied ☐
Neither satisfied nor dissatisfied ☐
Dissatisfied ☐
Very dissatisfied ☐

Q4 Overall, is this moisturizing cream
Better than your usual ☐
About the same ☐
Not as good as your usual ☐

Q5 What, if anything, did you LIKE about this moisturizing cream?

Q6 What, if anything, did you DISLIKE about this moisturizing cream?

Q7 How likely would you be to reuse this moisturizing cream?
I would definitely reuse it (skip Q8) ☐
I would probably reuse it (skip Q8) ☐
I would probably not reuse it (go to Q8) ☐
I would definitely not reuse it (go to Q8) ☐

Q8 Please explain why you would not be likely to reuse this moisturizing cream again?

Usage

Q9 The packaging of this moisturizing cream is
Very convenient ☐
Convenient ☐
Neither convenient nor inconvenient ☐
Inconvenient ☐
Very inconvenient ☐

Q10 Dispensing this moisturizing cream is
Very easy ☐
Easy ☐
Neither easy nor difficult ☐
Difficult ☐
Very difficult ☐

Q11 The consistency of this moisturizing cream is
Very thick ☐
Thick ☐

Neither thick nor thin ☐
Thin ☐
Very thin ☐

Q12 The consistency of this moisturizing cream is
Way too thick ☐
Too thick ☐
Just about right ☐
Too thin ☐
Way too thin ☐

Q13 The fragrance of this moisturizing cream is
Very pleasant ☐
Pleasant ☐
Neither pleasant nor ☐
unpleasant
Unpleasant ☐
Very unpleasant ☐

Q14 The fragrance of this moisturizing cream is
Way too intense ☐
Too intense ☐
Just about right ☐
Too weak ☐
Way too weak ☐

Application

Q15 Applying this moisturizing cream is
Very easy ☐
Easy ☐
Neither easy nor difficult ☐
Difficult ☐
Very difficult ☐

Q16 Absorption of this moisturizing cream is
Very fast ☐
Fast ☐
Neither quick nor slow ☐
Slow ☐
Very slow ☐

Q17 Absorption of this moisturizing cream is
Way too fast ☐
Too fast ☐
Just about right ☐
Too slow ☐
Way too slow ☐

Results

Q18 Once the cream is absorbed, your skin feels
Very hydrated ☐
Hydrated ☐
Neither hydrated nor dehydrated ☐
Dehydrated ☐
Very dehydrated ☐

Q19 Once the cream is absorbed, your skin feels
Way too greasy ☐
Too greasy ☐
Just about right ☐
Too dry ☐
Way too dry ☐

Q20 Once the cream is absorbed, your skin feels
Very smooth ☐
Smooth ☐
Neither smooth nor rough ☐
Rough ☐
Very rough ☐

End of day assessment

Q21 At the end of the day, your skin feels
Very hydrated ☐
Hydrated ☐
Neither hydrated nor dehydrated ☐
Dehydrated ☐
Very dehydrated ☐

Q22   At the end of the day, your skin feels
Way too greasy   ☐
Too greasy   ☐
Just about right   ☐
Too dry   ☐
Way too dry   ☐

Q23   At the end of the day, your skin feels
Very smooth   ☐
Smooth   ☐
Neither smooth nor rough ☐
Rough   ☐
Very rough   ☐

Depending on the objectives additional dimensions can be assessed.

Concept

Presentation of the concept to the consumer: could be via graphics, videos, audio, samples...

Q24   This concept is appealing.
Completely agree   ☐
Agree   ☐
Neither agree nor disagree ☐
Disagree   ☐
Completely disagree   ☐

Q25   This concept matches the moisturizing cream you just tested.
Completely agree   ☐
Agree   ☐
Neither agree nor disagree ☐
Disagree   ☐
Completely disagree   ☐

Section 1.2.2.2

Example of Recruitment Screener

This example is a scenario where the recruiter speaks directly with the potential recruit. It provides a variety of questions for different product categories.

Please introduce yourself and explain that the study requires a participation during a 4-weeks period. Make sure the person agrees on length and dates in case they qualify.

Q1 Please indicate the city where you live. (Quota)

Paris	☐	20%
Lyon	☐	20%
Marseille	☐	20%
Toulouse	☐	20%
Nancy	☐	20%

Q2 Please recruit only women (filter).

Male	☐	→Stop
Female	☐	

Q3 Could you please indicate your age? (Quota and filter)

Less than 18	☐	→Stop
18–24	☐	20%
25–34	☐	20%
35–44	☐	20%
45–54	☐	20%
55–64	☐	20%
65 or more	☐	→Stop

Q4 Do you or any of your close friends or family work in the areas below? (Filter areas that could present a conflict or a bias)

Advertising or PR	☐	→Stop
Journalism	☐	→Stop
Marketing	☐	→Stop
Bank	☐	
Insurance	☐	

... add multiple areas to avoid bias

Medicine/pharmacy	☐	→Stop (usually)
Telecommunications	☐	→Stop or continue depending on product field
Automobile industry	☐	→Stop or continue depending on product field
Personal care industry	☐	→Stop or continue depending on product field
Food industry	☐	→Stop or continue depending on product field

...

Beauty supply store	☐	→Stop or continue depending on product field
Department store	☐	→Stop or continue depending on product field
Grocery store/supermarket	☐	→Stop or continue depending on product field

Q5 Are you part of a consumer panel where you agree to test products or participate in surveys? (Filter)

Yes ☐ →Stop usually
No ☐

Q6 Have you participated in a product test over the past 6 months? (Filter)

... In some cases, answering yes on Q6 may be ok as long as it is not in the product test field category.

Yes ☐ →Stop usually
No ☐

Q7 Are you pregnant? (Filter)

Yes ☐ →Stop usually unless no product use is involved
No ☐

Q8 Have you ever experienced an allergic reaction? (Filter)

Yes ☐ →Stop usually, unless no product use is involved
No ☐

Q9 Are you taking any medications? (Filter)

Yes ☐ →Stop usually
No ☐

Q10 Among these social media, which ones are you familiar with? How often do you consult each one? (Filter and quota possible)

	Every day	5–6 times a week	3–4 times a week	Twice a week	Once a week	Less than once a week	Never
FaceTime					→Stop	→Stop	→Stop
Instagram							
Twitter							
Snapchat							
...							

Q11 You mentioned being familiar with FaceTime. Can you precise whether (filter)

You installed it yourself on your device ☐

You did not install it yourself but you ☐
would be able to do so if needed

You did not install it yourself and you ☐ →Stop
would not be able to do so without help

Q12 Where do you purchase your product? (Quota and filter)

Make sure to cover the country distribution channels.

Discount stores (Walmart, Target, etc.) ☐
Grocery stores (ShopRite, etc.) ☐
Beauty supply stores (Harmon, etc.) ☐
Whole sale clubs (Costco, etc.) ☐
Drugstores (CVS, etc.) ☐
Department stores (Macy's, etc.) ☐
Internet ☐
Other ☐

Q13 You mentioned using eyeliner. Can you precise which shade? (Quota and filter)

Black ☐ 50%
Blue ☐ 50%
Brown ☐ →Stop
Other ☐ →Stop

Q14 Can you precise which brands you use? (Quota and filter)

Question can be phrased in different ways depending on the target objective

Brand names	Check the brand you have used over the past 6 months	Check the brand you used the most often	Check your 'go to' brand
...			

Q15 For the brands you mentioned using over the past 6 months, could you precise how often you use each of them? (Quota and filter)

Brand used	Never	Rarely	Sometimes	Often	Very often
...					

Q16 What are the important criteria when you purchase your product? (Quota and filter)

Criteria	Not important	Somewhat important	Important	Very important	Extremely important
Brand					
Shade					
Smell					
...					

For some products, it may be necessary to screen hair length, colour, skin tones, types and so on. For individual in-depth interviews or focus groups, make sure that no one recruited is too embarrassed, inarticulate and that each participant converses clearly. It is also important for focus groups not to recruit close friends or relatives within the same group.

Section 1.2.2.2

Example of Consent Form

Below are listed standard statements typically included in a consent form for a sensory or consumer test. In all cases, it is always necessary to seek legal advice before finalizing and using a consent form.

Participant name:
Participant address:
Product(s) tested detailed name(s):

General information and agreement:

- The product(s) that I will try can be under development (not yet on the market) or already available to the general public.
- I agree to follow all instructions that are given to me.
- I have no sensitivity to any product to the best of my knowledge.
- I have no physical condition that could be adversely affected by the product(s) I will try.
- I am neither pregnant nor nursing.
- I agree not to take part in any other study during this one.

Consent: I consent to being photographed, filmed and interviewed by the employees of (Name of the market research company) and agree that any photos, videos and comments are the property of (Name of the market research company).

Privacy: I understand that data collected from this study will be used by (Name of the market research company) for product evaluation and marketing purposes. (Name of the market research company) will not release to the general public the information identifying me as required by law and for my safety. My identity may never be disclosed.

Confidentiality: I agree to maintain confidential all information about the product that I may learn. Products given to me are for my use only.

Volunteer participation: Participation is voluntary and I understand that I may refuse to participate and I may withdraw at any time without obligation or prejudice. I also understand that my participation may also be discontinued at any time without my consent by (Name of the market research company).

Contact information: I understand that if during test period, I experience non-serious medical problems or have any questions concerning the product, I can contact (contact information of the market research company). In case of an emergency, I understand that I should immediately seek medical attention.

If there can be any possible anticipated allergic reaction to the product, it needs to be disclosed.

Final statements:

I have had an opportunity to ask questions and my questions have been answered.

I have read and understood the above and agreed to this consent form.

If requested by me, I will be given a copy of this signed consent form.

I release (Name of the market research company) from any liability arising out of the use of the product being evaluated.

Signatures
Date:
Signature participant:
Name participant:

Date:
Signature witness:
Name witness:

Chapter 2

Section 2.3.1.1

Example of Sensory Questionnaire (for Coffee)

Oily

0	1	2	3	4	5	6	7	8	9	10
☐	☐	☐	☐	☐	☐	☐	☐	☐	☐	☐

Darkness

0	1	2	3	4	5	6	7	8	9	10
☐	☐	☐	☐	☐	☐	☐	☐	☐	☐	☐

Transparency

0	1	2	3	4	5	6	7	8	9	10
☐	☐	☐	☐	☐	☐	☐	☐	☐	☐	☐

Fragrance intensity

0	1	2	3	4	5	6	7	8	9	10
☐	☐	☐	☐	☐	☐	☐	☐	☐	☐	☐

Coffee typicality

0	1	2	3	4	5	6	7	8	9	10
☐	☐	☐	☐	☐	☐	☐	☐	☐	☐	☐

Acidity

0	1	2	3	4	5	6	7	8	9	10
☐	☐	☐	☐	☐	☐	☐	☐	☐	☐	☐

Bitterness

0	1	2	3	4	5	6	7	8	9	10
☐	☐	☐	☐	☐	☐	☐	☐	☐	☐	☐

Burnt

0	1	2	3	4	5	6	7	8	9	10
☐	☐	☐	☐	☐	☐	☐	☐	☐	☐	☐

Body

0	1	2	3	4	5	6	7	8	9	10
☐	☐	☐	☐	☐	☐	☐	☐	☐	☐	☐

Balance

0	1	2	3	4	5	6	7	8	9	10
☐	☐	☐	☐	☐	☐	☐	☐	☐	☐	☐

Arabica typicality

0	1	2	3	4	5	6	7	8	9	10
☐	☐	☐	☐	☐	☐	☐	☐	☐	☐	☐

Earthy

0	1	2	3	4	5	6	7	8	9	10
☐	☐	☐	☐	☐	☐	☐	☐	☐	☐	☐

Mouldy

0	1	2	3	4	5	6	7	8	9	10
☐	☐	☐	☐	☐	☐	☐	☐	☐	☐	☐

Persistency

0	1	2	3	4	5	6	7	8	9	10
☐	☐	☐	☐	☐	☐	☐	☐	☐	☐	☐

Chapter 3

Section 3.1.3a

Example of one-on-one discussion guide (1 h to go over test product, 1 h 30 min if detailed concept and/or packaging discussion)

Introduction (5 min)

- Moderator presentation
- Objectives of the interview: get your valuable input on the product tried at home
- State importance of honesty, sincerity, 'no right or wrong' attitude
- Participant's introduction: profession, family, and so forth.

Usual makeup routine (15 min)
Before going into detail on the product that you used at home, I would like to go over what you usually do.

- How often do you use makeup?
- Are they different circumstances where you use different products or routines?
- What products do you use? (Detail type and brand if possible.)
- How do you use them?
- What look are you trying to achieve? Is it different depending on circumstances?
- Are there some products you use for certain purposes and not for others?
- What do you like about the products you currently use (go over each product)?
- What do you dislike about the products you currently use (go over each product)?
- Is there anything you would wish for that you cannot currently find?
- Are there products you do not use anymore? Why?

In-depth discussion on test product (30 min)
Now, let us talk about the product you used at home.

Take me through the way you used the product over the week? Did you use it differently in the beginning and then over the course of the week? Did you make adjustments? If so, are they in the quantities used, in gesture or in placement during the routine?

- Take me through your detailed product experience
- What did you like about the product?
- What did you dislike about the product?

Probe on (depending on the type of product, interviewer can add details):

- Packaging (what did it make you think or expect?)
- Dispensing (easy to adjust quantities?)

- Consistency (likes/dislikes?)
- Application (easy/difficult?)
- Scent or flavour/fragrance (if applicable)
- Post application (right after)
- Look obtained (Able to achieve the desired look compared to other products?)
- How does the product last over the course of the day?

Wrap up/Round-Robin (10 min)

- What does the product do well?
- What are the aspects that should be improved?
- How does it compare to usual product(s)?

Concept exploration (15 min)
Example where objective is to build the concept.
Imagine you are requested to position the product.

- How would you describe the product?
- What does it do? What does it bring?
- How is it different from current products? How is it unique (if applicable)?
- What is its consumer target?
- In what context or circumstances should it be used?
- How would you recommend using it? When? To do or achieve what?

Packaging (15 min)
Example if need to explore packaging.

- What do you like about the packaging?
- What do you dislike about the packaging?
- Does it work?
- Would there be other ways to package it?
- What would these proposals bring?

Section 3.1.3b

Example of Focus groupdiscussion guide (2 h with prior product test, 1 h 30 min without prior product test)

Introduction (10 min)

- Moderator presentation
- Objectives of the discussion: presentation
- State importance of honesty, sincerity, 'no right or wrong' attitude
- Logistics: timing, breaks/restroom, food if applicable
- Participant's introduction round table

Individual routines and feel associated (35 min)
Objectives: Gather knowledge around daily routines, when women eat dairy products during the day, which products they choose at what point and why? Gather language used around the description and consumption of dairy products, understand 'healthy feel', how it is perceived and expressed, at what point during the dairy products consumption.
   One by one, each woman will address the different points below:

- How often do you eat dairy products? When during the day?
- What products (types, brands) do you eat at what point during the day? Why?

→*Moderator can list categories on an easel.*

Some products are consumed as standalones meals or snacks, some are considered as desserts, some may be used in recipes, and so on.

- Is a dairy product the first, intermediate or last product in your meal?
- Is it consumed alone?
- Do you mix other products with your dairy product, often, sometimes, very occasionally or never?
- If so, what additions do you like? Granola, fruit or any other? Why?
- Do you use it as an ingredient in recipes?
- What products can be an alternative to dairy products?
- How do they compare?

→*Moderator can list on an easel.*

- What is a good dairy product?
- What is a not-so-good dairy product?

→*Moderator can list categories on an easel.*

- What drives you to eat a dairy product?
- How do you feel when eating a dairy product?
- How do you feel after eating a dairy product?

Underline particular interest in those feelings.

- What is gained by consuming a dairy product?
- What are the good ways to feel?
- What are the mixed or not-so-good ways to feel
- How does it feel physically?
  - Good feel
  - Not-so-good feel
- How does it feel emotionally?
  - Good feel
  - Not-so-good feel

Going at list made on easel.

- In regard to mixed/not-so-good feels, how problematic are they?
- Are they associated particularly to certain types of products?

<u>Home product use (if applicable) (35 min)</u>
Let us talk about the product you tried at home.

- What did you like about it?
- What did you dislike about it?
- Is there anything unique about it compared to what you know?

  Let participants generate spontaneous comments as much as possible, and then probe if necessary:

- Visual aspect
- Consistency on spoon/in mouth
- Flavour in mouth
- After taste, if any
- Any other aspects that matter?

If multiple products, repeat for each product, and eventually ask for preference or ranking.

<u>Exploration of 'Healthy feel' (30 min)</u>
Discuss the concept

- When I say, 'Healthy feel', what does it mean for you?
- What other words mean the same or something similar?
- How important is it? How much does it matter?
- At what point during the consumption of a dairy product do you feel healthy?
- What are the elements that lead to feel healthy?
- What elements do not matter that much?

If product tested:

- Does the product emerge as healthy?
  - What aspects make you feel that way? (If healthy, what points matter; if unhealthy, what makes it fall behind?)
- If multiple products: Which product(s) emerge as the healthiest? (ranking or one product standing out) Why?
  - Compare preferred product versus healthiest, whether same or not, why? Understand to what extent the health factor is or not decisive.

<u>Wrap up/conclusion (10 min)</u>

- In new dairy products, what could a company do to really innovate? Is the healthy aspect the most important? What other aspects matter?
- Any questions?
- Thank you to all participants for their valuable input
- End

Section 3.2.2a

Example of Sniff Test Questionnaire

You will smell successively five different fragrances.
  Each fragrance is identified by a three-digit code.
  It is important that you smell the fragrances in the defined order.
  For each fragrance, please open the cap, sniff for a few seconds, then close the cap and fill the questions displayed for that sample.
  When filling the questionnaire, if you have a doubt, you can reopen the cap and sniff again. However, we recommend you always follow your initial impressions.
  Before moving on to the next sample, please sniff for a few seconds the coffee beans that are at your disposal to 'reset' your senses.
  Thank you very much for your input!

<u>Respondent identification</u>
Consumer code:
Sample code:

Q1 Overall, how much did you like this fragrance?

Disliked very much								Liked very much
1	2	3	4	5	6	7	8	9

Q2 How would you rate the intensity of this fragrance?

Too weak	Slightly too weak	Just about right	Slightly too strong	Too strong
1	2	3	4	5

Q3 This fragrance is…

	Fully disagree	Somewhat disagree	Neither agree nor disagree	Somewhat agree	Fully agree
Floral	1	2	3	4	5
Citrus	1	2	3	4	5
Woody	1	2	3	4	5

(Continued)

	Fully disagree	Somewhat disagree	Neither agree nor disagree	Somewhat agree	Fully agree
Oriental	1	2	3	4	5
Fruity	1	2	3	4	5
Green	1	2	3	4	5
Oceanic	1	2	3	4	5
Spicy	1	2	3	4	5

Q4 This fragrance is appropriate for...

	Fully disagree	Somewhat disagree	Neither agree nor disagree	Somewhat agree	Fully agree
Women	1	2	3	4	5
Men	1	2	3	4	5
Teens	1	2	3	4	5
Kids	1	2	3	4	5
The entire family	1	2	3	4	5
...					

Q5 This fragrance is appropriate for...

	Fully disagree	Somewhat disagree	Neither agree nor disagree	Somewhat agree	Fully agree
Hair care products	1	2	3	4	5
Laundry detergents	1	2	3	4	5
Room refresheners	1	2	3	4	5
A Perfume	1	2	3	4	5
An aftershave	1	2	3	4	5
...					

Q6 This fragrance makes me think of…

	Fully disagree	Somewhat disagree	Neither agree nor disagree	Somewhat agree	Fully agree
A cheap perfume	1	2	3	4	5
A cologne	1	2	3	4	5
A luxury perfume	1	2	3	4	5
…	1	2	3	4	5

Q7 This fragrance feels…

	Fully disagree	Somewhat disagree	Neither agree nor disagree	Somewhat agree	Fully agree
Soothing	1	2	3	4	5
Relaxing	1	2	3	4	5
Invigorating	1	2	3	4	5
Energizing	1	2	3	4	5
Stressful	1	2	3	4	5
…					

Section 3.2.2b

Example of Monadic Test Questionnaire

<u>Respondent identification</u>
Consumer code:
Period of use:

<u>Overall</u>

Q1 How many times did you use the pencil sharpener during the week?
Less than 5 times ☐
5 times ☐
6 times ☐
7 times ☐
More than 7 times ☐

Q2 Overall, how satisfied are you with the pencil sharpener?
Very satisfied ☐
Satisfied ☐
Neither satisfied nor dissatisfied ☐
Dissatisfied ☐
Very dissatisfied ☐

Q3 Overall, this pencil sharpener is...
Better than your usual ☐
About the same ☐
Not as good as your usual ☐

Q4 Overall, what, if anything, did you LIKE about the pencil sharpener you tested?

Q5 Overall, what, if anything, did you DISLIKE about the pencil sharpener you tested?

Q6 How likely would you be to reuse this pencil sharpener?
Definitely would reuse ☐
Probably would reuse ☐
Probably would not reuse ☐
Definitely would not reuse ☐

Usage

Q7 Holding this pencil sharpener is...
Very comfortable ☐
Comfortable ☐
Not very comfortable ☐
Not comfortable at all ☐

Q8 Sharpening a colour pencil with this sharpener is ...
Very easy ☐
Easy ☐
Neither easy nor difficult ☐
Difficult ☐
Very difficult ☐

Q9 The force you need to use is ...
Very strong ☐
Strong ☐
Not very strong ☐
Not strong at all ☐

Q10 For you, with this pencil sharpener, the force is …
Too strong ☐
Just right ☐
Not strong enough ☐

(…)

Results

Q11 Once sharpened, the tip of your colour pencil is …
Very sharp ☐
Sharp ☐
Not very sharp ☐
Not sharp at all ☐

Q12 Once sharpened, the tip of your colour pencil is …
Too sharp ☐
Just right ☐
Not sharp enough ☐

Q13 Once sharpened, the trim of the colour pencil tip is …
Very smooth ☐
Smooth ☐
Not very smooth ☐
Not smooth at all ☐

Q14 Once sharpened, the trim of the colour pencil tip is …
Too smooth ☐
Just right ☐
Not smooth enough ☐

Q15 The blade of the colour sharpener clogs …
Very easily ☐
Easily ☐
Not very easily ☐
Not easily at all ☐

(…)

Section 3.2.2c

Example of Sequential Monadic Questionnaire

The example given below shows the different types of scales that can be used: intensity, just-about-right, agreement.

You will test two different multipurpose spray cleaners for a period of 2 weeks each.

Each product is identified by a three-digit code.

It is important that you test the cleaners in the defined order. Thank you for being careful with the instructions regarding the codes and period of evaluation.

During the test period, you have to use the test cleaner ONLY (and not your usual cleaner).

Thank you for using each product at least 3 times during the 2-week period of test to allow you to have an accurate opinion.

You will be asked to fill out the questionnaire at the end of the 2-week period of use.

Thank you very much for your input!

Respondent identification
Consumer code:
Product code:
Period of use:

Usage details

Q1 How many times did you use the spray cleaner during the 2-week period?

Once ☐
Twice ☐
3 times ☐
4 times ☐
5 times or more ☐

Q2 Where did you use the spray cleaner in your home?

For bathrooms ☐
For kitchen ☐
For floor tiles ☐
For wood floors ☐
For furniture ☐
(...)

Q3 The days you used the spray cleaner, did you use it continuously for...

A few minutes ☐
15 min ☐
30 min ☐
1 h ☐
More than 1 h ☐

<u>Overall</u>

Q4   Overall, how much did you like the spray cleaner?

Disliked very much								Liked very much
1	2	3	4	5	6	7	8	9

Q5   Overall, how satisfied are you with the spray cleaner?
Very satisfied ☐
Satisfied ☐
Neither satisfied nor dissatisfied ☐
Dissatisfied ☐
Very dissatisfied ☐

Q6   Overall, the spray cleaner is…
Better than your usual ☐
About the same ☐
Not as good as your usual ☐

Q7   Overall, what, if anything, did you LIKE about the spray cleaner you tested?

Q8   Overall, what, if anything, did you DISLIKE about the spray cleaner you tested?

Q9   How likely would you be to reuse the spray cleaner?
Definitely would reuse ☐
Probably would reuse ☐
Probably would not reuse ☐
Definitely would not reuse ☐

Q10–Q16   Overall, how much did you like the spray cleaner specifically when…?

		Disliked very much								Liked very much
Q10	Taking it in your hand	1	2	3	4	5	6	7	8	9
Q11	Pressing the nozzle	1	2	3	4	5	6	7	8	9
Q12	Spraying it on a surface	1	2	3	4	5	6	7	8	9
Q13	Seeing the result	1	2	3	4	5	6	7	8	9
Q14	Smelling the fragrance	1	2	3	4	5	6	7	8	9

(Continued)

		Disliked very much								Liked very much
Q15	Seeing the cleaned surface once dried	1	2	3	4	5	6	7	8	9
Q16	Touching the cleaned surface once dried	1	2	3	4	5	6	7	8	9

Usage

Q17–Q24   Now focusing on when you use the spray cleaner, would you say...

		Not enough	Slightly not enough	Just about right	Slightly too much	Too much
Q17	Nozzle hard to press	1	2	3	4	5
Q18	Spray strength	1	2	3	4	5
Q19	Area covered by stream	1	2	3	4	5
Q20	Size of the droplets	1	2	3	4	5
Q21	Intensity of the fragrance during usage	1	2	3	4	5
Q22	Wetting power	1	2	3	4	5
Q23	Drying time	1	2	3	4	5
Q24	Intensity of the fragrance once dried	1	2	3	4	5

Q25–Q36   Still focusing on the usage of the spray cleaner, please indicate how much you agree or disagree with the following statements.

		Fully disagree	Somewhat disagree	Neither agree nor disagree	Somewhat agree	Fully agree
Q25	The package has a good fit in hand	1	2	3	4	5
Q26	The package is sturdy	1	2	3	4	5

(Continued)

		Fully disagree	Somewhat disagree	Neither agree nor disagree	Somewhat agree	Fully agree
Q27	The cleaner can be carried easily	1	2	3	4	5
Q28	The cleaner can be used everywhere in the house	1	2	3	4	5
Q29	The nozzle is easy to press	1	2	3	4	5
Q30	The nozzle fits to the finger	1	2	3	4	5
Q31	The nozzle is free of residue once used	1	2	3	4	5
Q32	It is easy to control the stream	1	2	3	4	5
Q33	The nozzle does not clog	1	2	3	4	5
Q34	The fragrance is nor harsh	1	2	3	4	5
Q35	The spray distributes evenly on the surface	1	2	3	4	5
Q36	The spray is fast to dry	1	2	3	4	5

Results

Q37–Q40 Now focusing on results once you are done using the cleaner, would you say...

		Fully disagree	Somewhat disagree	Neither agree nor disagree	Somewhat agree	Fully agree
Q37	The surface is evenly clean	1	2	3	4	5

(Continued)

		Fully disagree	Somewhat disagree	Neither agree nor disagree	Somewhat agree	Fully agree
Q38	The surface is evenly shiny	1	2	3	4	5
Q39	The fragrance left is not harsh	1	2	3	4	5
Q40	The surface is not slippery	1	2	3	4	5

Questions can in some cases be reiterated at multiple successive time points if relevant (during usage, right after usage, after a few hours, at the end of the day, etc.)

After the second product questionnaire

Q41   Which spray cleaner do you prefer?
Spray cleaner #256   ☐
Spray cleaner #478   ☐
Both   ☐
Neither   ☐

Q42   Reasons for preference (check all that apply)
Convenience   ☐
Nozzle   ☐
Spray force   ☐
Spray pattern   ☐
Spray distribution   ☐
Fragrance   ☐
Drying time   ☐
Other   ☐

Section 3.3.3

Example of discussion guide for a home visit (1 h 30 min in this example)

Introduction (15 min)

- Interviewer presentation
- Objectives of the visit: understand usage and sentiment on digital media players

- State importance of honesty, sincerity, 'no right or wrong' attitude
- Participant introduction: name, family/living situation, profession.
- Service used and most recent apps used (always the same or likes to explore new ones, etc.)

Take pictures of the environment and setting.

Follow every step of the usage experience (30 min)

I am interested in understanding how you use your digital media player, what apps you have, wish to have and how you feel at every step of the usage experience. Please narrate everything that you feel and comes to your mind as you move along the process. Take me to the usage of your digital media player as you would do if I was not here!

Interviewer takes notes and reimpulses the narration if necessary.

Postusage interview (30 min)

Overall

- What do you like about this service?
- What do you dislike about it?

Specifics (as needed)

- Turning the system on
  - Positive/negative/neutral?
  - Compared to other TV options?
- Switching from one app to another
  - Positive/negative/neutral?
  - Compared to other TV options?
- Your remote control
  - Positive/negative/neutral?
  - Compared to other TV options?
- Downloading a new app
  - Describe
  - Positive/negative/neutral?
  - Compared to other TV options?
    (...)

Wrap up (15 min)

To sum up, the reasons you like this service and will you continue using it?
Areas to be improved?
Ideal environment and options?
Elements you do not have today that you wish you had?
Thanks very much!

Index